W9-DBZ-738

Dr. Anton Reiner
Department of Anatomy and Neurobiology
The University of Tennessee Memphis
The Health Science Center
875 Monroe Avenue
Memphis, Tn 38163

PHARMACOLOGICAL REGULATION *of* GENE EXPRESSION *in the* CNS

Pharmacology and Toxicology: Basic and Clinical Aspects

Mannfred A. Hollinger, Series Editor
University of California, Davis

Forthcoming Titles

Antibody Therapeutics, William J. Harris and John R. Adair
Anabolic Treatment for Osteoporosis, James F. Whitfield and Paul Morley
Antisense Oligodeonucleotides as Novel Pharmacological Therapeutic Agents, Benjamin Weiss
Basis to Toxicity Testing, Second Edition, Donald J. Ecobichon
CNS Injuries: Cellular Responses and Pharmacological Strategies, Martin Berry and Ann Logan
Lead and Public Health: Integrated Risk Assessment, Paul Mushak
Molecular Bases of Anesthesia, Eric Moody and Phil Skolnick
Muscarinic Receptor Subtypes in Smooth Muscle, Richard M. Eglen
Receptor Characterization and Regulation, Devendra K. Agrawal

Published Titles

Inflammatory Cells and Mediators in Bronchial Asthma, 1990, Devendra K. Agrawal and Robert G. Townley
Pharmacology of the Skin, 1991, Hasan Mukhtar
In Vitro *Methods of Toxicology*, 1992, Ronald R. Watson
Basis of Toxicity Testing, 1992, Donald J. Ecobichon
Human Drug Metabolism from Molecular Biology to Man, 1992, Elizabeth Jeffreys
Platelet Activating Factor Receptor: Signal Mechanisms and Molecular Biology, 1992, Shivendra D. Shukla
Biopharmaceutics of Ocular Drug Delivery, 1992, Peter Edman
Beneficial and Toxic Effects of Aspirin, 1993, Susan E. Feinman
Preclinical and Clinical Modulation of Anticancer Drugs, 1993, Kenneth D. Tew, Peter Houghton, and Janet Houghton
Peroxisome Proliferators: Unique Inducers of Drug-Metabolizing Enzymes, 1994, David E. Moody
Angiotensin II Receptors, Volume I: Molecular Biology, Biochemistry, Pharmacology, and Clinical Perspectives, 1994, Robert R. Ruffolo, Jr.
Angiotensin II Receptors, Volume II: Medicinal Chemistry, 1994, Robert R. Ruffolo, Jr.
Chemical and Structural Approaches to Rational Drug Design, 1994, David B. Weiner and William V. Williams
Biological Approaches to Rational Drug Design, 1994, David B. Weiner and William V. Williams

Published Titles Continued

Inflammatory Cells and Mediators in Bronchial Asthma, 1990,
Direct Allosteric Control of Glutamate Receptors, 1994, M. Palfreyman,
 I. Reynolds, and P. Skolnick
Genomic and Non-Genomic Effects of Aldosterone, 1994, Martin Wehling
Human Growth Hormone Pharmacology: Basic and Clinical Aspects, 1995,
 Kathleen T. Shiverick and Arlan Rosenbloom
Placental Toxicology, 1995, B. V. Rama Sastry
Stealth Liposomes, 1995, Danilo Lasic and Frank Martin
TAXOL®: Science and Applications, 1995, Matthew Suffness
Endothelin Receptors: From the Gene to the Human, 1995, Robert R. Ruffolo, Jr.
Alternative Methodologies for the Safety Evaluation of Chemicals
 in the Cosmetic Industry, 1995, Nicola Loprieno
Phospholipase A$_2$ in Clinical Inflammation: Molecular Approaches
 to Pathophysiology, 1995, Keith B. Glaser and Peter Vadas
Serotonin and Gastrointestinal Function, 1995, Timothy S. Gaginella and
 James J. Galligan
Drug Delivery Systems, 1996, Vasant V. Ranade and Mannfred A. Hollinger
Experimental Models of Mucosal Inflammation, 1996, Timothy S. Gaginella
Brain Mechanisms and Psychotropic Drugs, 1996, Andrius Baskys and
 Gary Remington
Receptor Dynamics in Neural Development, 1996, Christopher A. Shaw
Ryanodine Receptors, 1996, Vincenzo Sorrentino
Therapeutic Modulation of Cytokines, 1996, M.W. Bodmer and Brian Henderson
Pharmacology in Exercise and Sport, 1996, Satu M. Somani
Placental Pharmacology, 1996, B. V. Rama Sastry
Pharmacological Effects of Ethanol on the Nervous System, 1996,
 Richard A. Deitrich
Immunopharmaceuticals, 1996, Edward S. Kimball
Chemoattractant Ligands and Their Receptors, 1996, Richard Horuk
Pharmacological Regulation of Gene Expression in the CNS,
 1996, Kalpana Merchant

PHARMACOLOGICAL REGULATION *of* GENE EXPRESSION *in the* CNS

Edited by

Kalpana Merchant, Ph.D.

CNS Diseases Research
Pharmacia & Upjohn, Inc.
Kalamazoo, Michigan

CRC Press
Boca Raton   New York   London   Tokyo

Acquiring Editor: Paul Petralia
Project Editor: Renee Taub
Marketing Manager: Susie Carlisle
Direct Marketing Manager: Becky McEldowney
Cover design: Dawn Boyd
PrePress: Carlos Esser
Manufacturing: Sheri Schwartz

Library of Congress Cataloging-in-Publication Data

Pharmacological regulation of gene expression in the CNS / edited by
 Kalpana Merchant.
        p.   cm. -- (Pharmacology and toxicology)
     Includes bibliographical references and index.
     ISBN 0-8493-8550-4 (alk. paper)
     1. Neostriatum--Physiology.   2. Gene expression--Regulation.
 3. Dopamine--Physiological effect.   4. Dopamine--Receptors.
 I. Merchant, Kalpana.   II. Series: Pharmacology & toxicology (Boca
 Raton, Fla.)
     [DNLM: 1. Brain--physiology.   2. Neostriatum--metabolism.   3. Gene
 Expression Regulation--drug effects.   4. Receptors, Dopamine-
 -genetics.   WL 307 P536 1996]
 QP383.9.P48   1996
 612.8'25--dc20
 DNLM/DLC
 for Library of Congress   96-18410
                                                                          CIP

     This book contains information obtained from authentic and highly regarded sources.
Reprinted material is quoted with permission, and sources are indicated. A wide variety of
references are listed. Reasonable efforts have been made to publish reliable data and information,
but the author and the publisher cannot assume responsibility for the validity of all materials or
for the consequences of their use.

     Neither this book nor any part may be reproduced or transmitted in any form or by any means,
electronic or mechanical, including photocopying, microfilming, and recording, or by any information storage or retrieval system, without prior permission in writing from the publisher.
     All rights reserved. Authorization to photocopy items for internal or personal use, or the
personal or internal use of specific clients, may be granted by CRC Press, Inc., provided that $.50
per page photocopied is paid directly to Copyright Clearance Center, 27 Congress Street, Salem,
MA 01970 USA. The fee code for users of the Transactional Reporting Service is ISBN 0-8493-
8550-4/96/$0.00+$.50. The fee is subject to change without notice. For organizations that have
been granted a photocopy license by the CCC, a separate system of payment has been arranged.
     CRC Press, Inc.'s consent does not extend to copying for general distribution, for promotion,
for creating new works, or for resale. Specific permission must be obtained in writing from CRC
Press for such copying.
     Direct all inquiries to CRC Press, Inc., 2000 Corporate Blvd., N.W., Boca Raton, Florida
33431.

© 1996 by CRC Press, Inc.

No claim to original U.S. Government works
International Standard Book Number 0-8493-8550-4
Library of Congress Card Number 96-18410
Printed in the United States of America 1 2 3 4 5 6 7 8 9 0
Printed on acid-free paper

PREFACE

Extracellular environmental signals received by neurons first produce an activation of second messenger systems associated with cell surface receptors. This kind of an immediate biological response is often short-lived, governed primarily by temporal patterns of protein phosphorylation/dephosphorylation events. However, it can lead rapidly to specific short- and long-term changes in gene expression which, in some cases, may permanently alter the phenotype of a neuron. Steroid hormone receptors can bypass the second messenger systems and directly lead to genomic responses. The phenotypic alterations in the neurons determine the outcome of subsequent homologous and heterologous cell surface stimuli. This phenomenon, termed neuronal plasticity, plays a critical role in physiological as well as pathological effects in the central nervous system. Learning and memory is a prime example of a physiological process that involves genomic responses. On the other hand, stress-induced alterations in neuronal phenotype may underlie the pathophysiology of diseases such as schizophrenia and depression. It would be obvious then that phenotypic adaptations by neurons to pharmacological agents play a pivotal role not only in dictating the immediate response but also in such phenomena as drug tolerance and sensitization thought to underlie dependence and addiction liability.

The importance of neuronal plasticity in pharmacological regulation of neuronal activity cannot be exemplified better than that demonstrated by studies of dopamine-mediated gene regulation in the neostriatum. The neostriatum is a neurochemically rich structure receiving converging neural inputs from a number of distant sites associated with the basal ganglia. The integration of various neural signals within the neostriatum regulates such complex behaviors as emotions, motivation, addiction as well as movement, and dopaminergic neurotransmission plays a key role in this process. Alterations in dopamine neurotransmission are implicated in the pathophysiology and pharmacotherapy of psychotic disorders (such as schizophrenia) and movement disorder (e.g., Parkinson's disease), as well as addiction to psychostimulants such as cocaine and amphetamine. The efficacy of, as well as clinical complications produced by, long-term dopaminergic therapy of these diseases involves adaptive alterations in targeted neurons. Additionally, the addiction liability of psychostimulants is thought to be due to phenotypic alterations induced by endogenously released dopamine. Hence, an understanding of the influence of dopamine in acute and long-term changes in gene expression within the neostriatum is crucial for gaining insight into cellular mechanisms involved in integration of divergent stimuli within the striatum and in phenomena involved in neuronal adaptations. This in turn will shed light on the pathophysiology of central disorders involving the basal ganglia and offer fundamentally novel approaches for the treatment of these diseases.

This volume is planned to provide a comprehensive overview of dopamine-mediated regulation of gene expression within the striatum and associated regions. The contributors have reviewed a range of studies discussing the interactions of major dopamine receptor families in regulating acute and chronic gene expression within the intrinsic, afferent, and efferent striatal neurons. Additionally, in view of the now established heterogeneity of dopamine receptor subtypes within the major families, a potential role of each receptor subtype in mediating phenotypic and behavioral effects is discussed. The results were derived from the effects of subtype-specific agents as well as the approach of selective knock-down of specific receptors with antisense oligonucleotides. Studies of coordinated expression of transcription regulatory factors, neuropeptide genes as well as genes involved in the synthesis of classical neurotransmitters provide an insight into the complexity of dynamic interactions between transmitter systems within the neostriatum. These articles also provide proof that pharmacological alteration in gene expression is an effective method to understand neurotransmitter interactions, both at the level of neural systems and cellular level. Finally, the correlation between drug-induced, long-term phenotypic changes with behavioral effects and the utility of this approach for predicting clinical outcome is exemplified in chronic studies of antipsychotic and antiparkinsonian agents.

I would like to acknowledge my sincere gratitude to all the contributors of this monograph for sharing my enthusiasm for this volume and making time from their extremely busy schedules for careful preparation of each chapter. Most of all, I am indebted to my husband, Mahesh, for his unending encouragement, constructive criticism, and tolerance of my labors. I would not have taken on this daunting task without his support and so to him I dedicate this book with affection.

Kalpana M. Merchant

THE EDITOR

Kalpana M. Merchant, Ph.D., is a Research Scientist in the CNS Diseases Research at the Pharmacia & Upjohn, Inc., in Kalamazoo, Michigan.

Dr. Merchant received her B.S. in Pharmacy from Bombay University (India) in 1979. After obtaining research experience at multinational pharmaceutical companies in India for 5 years, she joined the Department of Pharmacology and Toxicology at the University of Utah and obtained her Ph.D. in 1989. She did a postdoctoral fellowship in the Department of Pharmacology at the University of Washington in Seattle for 2 years after which she was appointed first as an Acting Assistant Professor and then as Assistant Professor in the Department of Psychiatry at the University of Washington, Seattle, in 1991. In 1993 she joined The Upjohn Company as a Research Scientist with an Adjunct appointment in the Department of Biological Sciences at Western Michigan University.

Dr. Merchant is a member of the Society for Neuroscience, the American Association for the Advancement of Sciences, International Neuropeptide Society, and Indian Pharmaceutical Association.

She received several undergraduate and graduate scholarships and has been a recipient of the Young Investigator Award from the International Schizophrenia Research program. Dr. Merchant received research grants from the Washington Institute for Mental Health Research and Training, The Scottish Rite Schizophrenia Foundation, The Stanley Research Foundation and The National Institute for Neurological Disorders and Stroke. Her current major research interests include molecular mechanisms underlying the pathophysiology and pharmacotherapy of diseases involving central dopamine systems. Specifically, she studies the involvement of transcription factors and their downstream genetic targets in the pharmacological effects of psychoactive drugs and in animal models of diseases like schizophrenia and Parkinson's disease.

CONTRIBUTORS

Marie-Francoise Chesselet
Department of Pharmacology
University of Pennsylvania
School of Medicine
Philadelphia, Pennsylvania

Douglas Cole
Laboratory of Molecular and
*   Developmental Neuroscience*
Massachusetts General Hospital
Charlestown, Massachusetts

Rebecca L. Cole
Laboratory of Molecular and
*   Developmental Neuroscience*
Charlestown, Massachusetts

Ian Creese
Center for Molecular and Behavioral
*   Neuroscience*
Rutgers, the State University
*   of New Jersey*
Newark, New Jersey

Genoveva Davidkova
Department of Pharmacology
Medical College of Pennsylvania
Philadelphia, Pennsylvania

Jill M. Delfs
Department of Pharmacology
University of Pennsylvania
School of Medicine
Philadelphia, Pennsylvania

Charles R. Gerfen
Laboratory of Neurophysiology
National Institutes of Mental Health
Bethesda, Maryland

Bruce Hope
Laboratory of Molecular and
*   Developmental Neuroscience*
Massachusetts General Hospital
Charlestown, Massachusetts

Steven E. Hyman
Laboratory of Molecular and
*   Developmental Neuroscience*
Massachusetts General Hospital
Charlestown, Massachusetts

Kristen A. Keefe
Laboratory of Neurophysiology
National Institutes of Mental Health
Bethesda, Maryland

Christine Konradi
Laboratory of Molecular and
*   Developmental Neuroscience*
Massachusetts General Hospital
Charlestown, Massachusetts

Lynn P. Martin
Center for Molecular and
*   Behavioral Neuroscience*
Rutgers, the State University
*   of New Jersey*
Newark, New Jersey

Jacqueline F. McGinty
Department of Anatomy and Cell Biology
East Carolina University School of
Medicine
Greenville, North Carolina

Kalpana Merchant
CNS Diseases Research
Pharmacia & Upjohn, Inc.
Kalamazoo, Michigan

Abdel-Mouttalib Ouagazzal
Center for Molecular and
Behavioral Neuroscience
Rutgers, the State University
of New Jersey
Newark, New Jersey

George S. Robertson
Department of Pharmacology
University of Ottawa
Ottawa, Ontario, Canada

Michael Schwarzchild
Laboratory of Molecular and
Developmental Neuroscience
Massachusetts General Hospital
Charlestown, Massachusetts

Heinz Steiner
Laboratory of Neurophysiology
National Institutes of Mental Health
Bethesda, Maryland

Bao-Cun Sun
Center for Molecular and
Behavioral Neuroscience
Rutgers, the State University
of New Jersey
Newark, New Jersey

James M. Tepper
Center for Molecular and
Behavioral Neuroscience
Rutgers, the State University
of New Jersey
Newark, New Jersey

John Q. Wang
Department of Anatomy and Cell Biology
East Carolina University
School of Medicine
Greenville, North Carolina

Benjamin Weiss
Department of Pharmacology
Medical College of Pennsylvania
Philadelphia, Pennsylvania

Ming Zhang
Center for Molecular and Behavioral
Neuroscience
Rutgers, the State University
of New Jersey
Newark, New Jersey

Long-Wu Zhou
Department of Pharmacology
Medical College of Pennsylvania
Philadelphia, Pennsylvania

TABLE OF CONTENTS

PART III PERSISTENT ALTERATIONS IN GENE EXPRESSION INVOLVED IN BEHAVIORAL ADAPTATIONS

Part I

Dopamine-Mediated Gene Expression in the Neostriatum

Chapter 1

D1 AND D2 DOPAMINE RECEPTOR-MEDIATED GENE REGULATION IN THE STRIATUM

Charles R. Gerfen, Kristen A. Keefe and Heinz Steiner

CONTENTS

1. INTRODUCTION

Dopamine-containing neurons in the midbrain provide a massive input to the striatum,[1] the principal nucleus of the basal ganglia. Evidence of the critical role of this neurotransmitter in the striatum comes from the clinical disorders that result from the neurodegenerative loss of striatal dopamine in Parkinson's disease[2] and the implication of striatal dopamine effects of drugs of abuse including cocaine and amphetamine. The dynamic response of a number of different genes to manipulation of dopamine receptor activation has provided the basis of a strategy for analysis of dopamine's functional role in the basal ganglia. Altered levels of messenger RNAs encoding a variety of proteins and peptides within striatal neurons in response to pharmacologic treatments may be measured with *in situ* hybridization histochemical techniques. Such analysis provides the ability to quantify dopamine receptor-mediated effects in subsets of striatal neurons that are identified on the basis of their connections and/or receptor subtype phenotype. What has emerged from such studies is a view of the dynamic modulatory role that dopamine plays in affecting basal ganglia function. First, the connectional organization of the basal ganglia displays the existence of separate striatal output systems that have antagonistic effects on basal ganglia output.[2,3] The two major families of dopamine receptors, the D1 and D2 subtypes,[4] are differentially expressed by the two striatal output systems.[5] Second, the neuropeptides that co-segregate in neurons with the D1 and D2 receptor subtypes are oppositely affected by activation of these receptors.[5] Thus, these neuropeptides provide diagnostic measures of the effects of activation of dopamine receptors on functionally identified striatal neuron populations. Third, immediate early gene transcription factors display rapid responses to dopamine receptor activation that provide further demonstration of the opposite effects of D1 and D2 receptor activation on striatal output systems.[6-8] Moreover, they also demonstrate intercellular interactions between striatal neurons in response to dopamine receptor manipulation.[8] Fourth, repeated administration of cocaine alters striatal dynorphin levels which results in diminished D1 dopamine receptor-mediated immediate early gene induction in the striatum in response to cocaine.[9,10] Results indicate regional variations in such effects and suggest possible adaptive mechanisms within the striatum linking opioid and dopamine receptor modulation of basal ganglia function. Together these findings suggest that dopamine modulates the relative activity of the two striatal output systems which in turn determines the output of the basal ganglia, which feeds back to the frontal cortical areas involved in the preparatory functions of behavior.

2. FUNCTIONAL ORGANIZATION OF THE BASAL GANGLIA

Consideration of dopamine's function in the striatum requires a brief overview of the functional organization of the basal ganglia and its major nuclear complex, the striatum (Figure 1 and for review see Gerfen, 1992[3] and Gerfen and Wilson, 1996 [11]). The striatum, which is composed of the caudate nucleus, putamen, and nucleus accumbens, receives excitatory input from most of the cerebral cortex. This input is organized both topographically and convergently, such that there is an overlap of projections into the striatum from cortical areas that are interconnected. Cortical input targets the major neuron cell type in the striatum, the spiny projection neuron,[12] which comprises over 90% of the neurons in the striatum. Spiny projection neurons are composed of two main types,[13] those which project directly and those which project indirectly to the GABAergic output neurons of the basal ganglia in the entopeduncular nucleus (internal segment of the globus pallidus in primates) and the substantia nigra pars reticulata. The output neurons of the basal ganglia provide a tonic inhibition to thalamic nuclei that project to frontal cortical areas and to the superior colliculus and pedunculopontine nucleus. Striatal regulation of this inhibitory output is determined by the relative activity of the direct and indirect striatal output pathways. The direct pathway provides inhibitory input, which serves to disinhibit and thus facilitate activity in the thalamocortical connections,[14] whereas the indirect pathway, by way of connections through the globus pallidus and subthalamic nucleus provides excitatory input,[15,16] and thus serves to increase inhibition of thalamocortical connections. While this summary of the functional organization of the striatum is oversimplified, it serves as a model for examining dopamine function in the striatum. Moreover, despite its oversimplification, the model appears to provide a reasonable explanation for the clinical akinesia of Parkinson's disease. According to the model, dopamine depletion leads to increased output of the indirect striatal output system, which due to the resultant increased inhibition of thalamocortical connections, is responsible for Parkinsonian akinesia.[2] Reversal of Parkinsonian akinesia in both animal models and human clinical cases with lesions that target the indirect striatal output system provide support for the validity of the model.[17]

3. NEUROCHEMICAL PHENOTYPE OF STRIATAL OUTPUT NEURONS

As stated, spiny projection neurons, which constitute over 90% of the striatal neuron population, may be subdivided into two types based

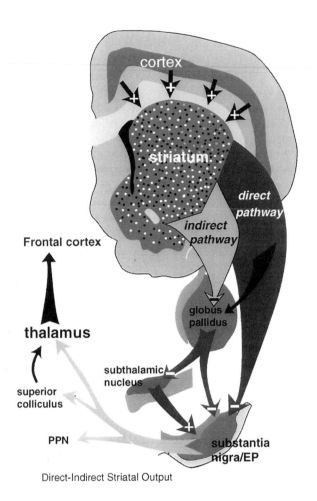

Direct-Indirect Striatal Output

FIGURE 1.
Summary diagram of the "direct" and "indirect" striatal output pathways. Layer 5 cortical neurons provide excitatory input (+) to the striatum. The direct striatal projection is provided by D1/substance P/dynorphin-containing neurons to the substantia nigra and entopeduncular nucleus and, to a lesser degree, to the globus pallidus. The indirect striatal projection is provided by D2/enkephalin-containing neurons that project to the globus pallidus. The globus pallidus in turn provides an inhibitory projection to the substantia nigra and to the subthalamic nucleus. The subthalamic nucleus provides an excitatory input to the substantia nigra. Thus, the direct and indirect pathways provide antagonistic input to the substantia nigra. The GABA neurons in the substantia nigra provide an inhibitory projection to the superior colliculus, pedunculopontine nucleus (PPN; not shown) and thalamus. The thalamic nuclei receiving this output project back upon the frontal cortex. The entopeduncular (EP) nucleus is connected in a similar manner to the substantia nigra but is not shown.

on their projection targets,[13] which are of roughly equal numbers.[18] One type, referred to as direct-projecting neurons, extends an axon collateral into the globus pallidus, and other collaterals to the entopeduncular nucleus and/or the substantia nigra. The designation of these neurons is based on their direct projection to the output nuclei of the basal ganglia. The second type of neuron, referred to as indirect-projecting neurons, provides an axon that projects to and arborizes extensively within the globus pallidus. The designation of these neurons is based on the multisynaptic connections of these neurons with the output neurons of the basal ganglia, which include globus pallidus to substantia nigra connections and globus pallidus to subthalamic nucleus to substantia nigra connections. Direct and indirect neurons thus are connectionally distinct striatal neuron populations. Estimates of the percentage of these two types of projection types approximates 50% for each.[18]

Both direct and indirect spiny projection neurons utilize GABA as their primary neurotransmitter,[19] and thus provide inhibitory inputs to their targets.[14] However, these two neuron populations are distinct in their differential expression of a number of other neurochemical markers (Figure 2). Direct-projecting neurons express the D1 dopamine receptor subtype[5] and the peptides substance P and dynorphin.[18] Indirect-projecting neurons express the D2 dopamine receptor subtype[5] and the peptide enkephalin.[18] Although these relationships are certainly not absolute, they are remarkably accurate approximations. For example, it has been reported that greater than 95% of D1 containing neurons co-localize substance P mRNA, whereas greater than 95% of D2 containing neurons co-express enkephalin mRNA.[20]

4. DOPAMINE-MEDIATED PEPTIDE GENE REGULATION

Early studies by several groups reported changes in levels of striatal peptides following manipulation of dopamine neurotransmission. Treatment of animals with neuroleptics, D2 receptor antagonists, were reported to result in the elevation of enkephalin peptide levels[21] (see also Chapters 2 and 7). Treatment with psychostimulants, such as amphetamine, resulted in elevated dynorphin levels[22] (see also Chapter 4). The advent of the use of *in situ* hybridization histochemical localization of the mRNAs encoding the precursors of enkephalin, substance P and dynorphin provided the ability to analyze dopamine-mediated changes in the levels of these peptides at the cellular level. Using this technique it was reported that following dopamine depletion of the striatum enkephalin mRNA levels increased, whereas substance P mRNA levels decreased,[23] consistent with prior studies.

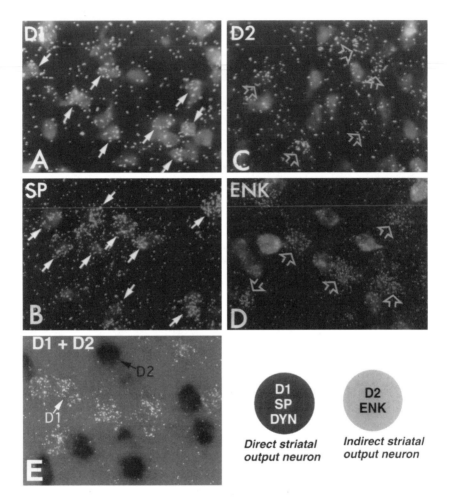

FIGURE 2.

In situ hybridization histochemical localization of mRNAs to identify peptides and dopamine receptor subtypes in striatal spiny projection neurons. Striatonigral neurons contain both D1 and substance P mRNAs, whereas striatopallidal neurons contain both D2 and enkephalin mRNAs. (A-C) Neurons that project to the substantia nigra are retrogradely labeled with the fluorescent dye fluorogold (black labeled cell bodies). *In situ* hybridization labeling of mRNA is shown by white grains. (A) D1 dopamine receptor mRNA is localized in labeled striatonigral neurons (arrows). (B) Substance P mRNA is also localized in labeled striatonigral neurons (arrows). (C) D2 dopamine receptor mRNA is not contained in labeled striatonigral neurons but in unlabeled striatopallidal neurons (open arrows). (D) Enkephalin mRNA is also contained in unlabeled striatopallidal neurons (open arrows). (E) Both D1 and D2 mRNAs are labeled in the same section, D1 mRNA with an S35-riboprobe that is marked by white silver grains over neurons and D2 mRNA with a digoxigenin-riboprobe that is labeled with a dark immunoreactive reaction. D1 and D2 mRNAs are segregated in separate neurons, with less than 5% of the entire population of striatal spiny projection neurons containing appreciable amounts of both receptor subtypes. ([A-C] are from Gerfen, C. R. et al., *Science*, 250, 1429, 1990.)

In situ hybridization methods provide an additional level of analysis in the ability to more clearly characterize the neurochemical phenotype of striatal neuron populations. Early studies on the segregation of enkephalin in the indirect pathway and substance P and dynorphin in the direct pathway used immunohistochemical techniques.[24,25] These techniques were best at localization of peptide immunoreactivity in fibers and terminals, which showed greater enkephalin staining in the target area of the indirect pathway, the globus pallidus, and greater staining of substance P and dynorphin in the target area of the direct pathway, the substantia nigra. However, these methods provided less than optimal labeling of the neurons of origin in the striatum, which was subject to fixation variables.[26] Combining fluorescent retrograde tracing with *in situ* hybridization histochemical localization of mRNAs provided a direct demonstration that indirect-projecting neurons express enkephalin and direct projecting neurons express both substance P and dynorphin[18] (see also Chapter 6). This study established that the two pathways were composed of roughly equal numbers of neurons, that each striatal projection subtype was evenly distributed throughout the striatum in both patch and matrix compartments, and that they were intermingled with each other. Further quantitative study at the single-cell level of dopamine-mediated changes in peptide mRNA levels not only confirmed that dopamine depletion resulted in elevated enkephalin and decreased substance P and dynorphin levels, and that dopamine agonist treatment elevated substance P and dynorphin, but added that the changes in peptide mRNA levels did not occur as a change in the peptide phenotype of the neurons.[27] This finding was critical in that it established the segregation of enkephalin to indirect and substance P and dynorphin to direct striatal projection neurons, such that changes in these neuropeptides are diagnostic of pathway specific, dopamine-mediated effects.

The next stage in understanding dopamine-mediated regulation of striatal function came from the determination that the D1 and D2 dopamine receptor subtypes are, respectively, segregated along with dynorphin and substance P in direct-projecting neurons and with enkephalin in indirect-projecting neurons.[5] This was established both with combined *in situ* hybridization in connectionally identified neurons[5] and by co-localization of the receptor subtype and peptide mRNAs.[20,28,29] The development of D1 and D2 receptor specific antibodies has enabled the localization of the actual receptor proteins at the cellular and ultrastructural level in the striatum.[30] Such studies confirm the segregation of D1 and D2 receptors to separate populations of striatal projection neurons.[31]

The segregation of D1 receptors to direct and D2 receptors to indirect striatal projection neurons is consistent with the opposite effects of dopamine depletion on peptide levels in these neurons (Figure 3).

control striatum	6-OHDA lesion	6-OHDA + D1 agonist	6-OHDA + D2 agonist

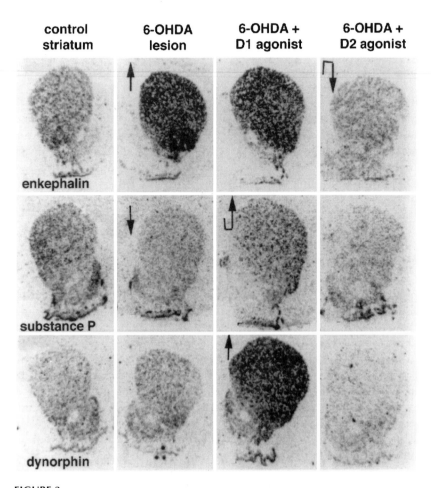

FIGURE 3.

Data from an experiment demonstrating D1- and D2-receptor selective gene regulation in direct and indirect striatal projection neurons. Images are of coronal sections of film autoradiographs labeled with *in situ* hybridization histochemical localization of enkephalin mRNA (top row), substance P mRNA (middle row) and dynorphin mRNA (bottom row) from an intact rat striatum (control striatum: first column), from a striatum depleted of dopamine (6-OHDA lesion: second column), from a dopamine-depleted striatum of an animal treated with the D1 agonist SKF 38393 (single daily injections 12.5 mg/kg for 21 days) and from a dopamine-depleted striatum of an animal treated with a D2 agonist quinpirole (continuous treatment of 1 mg/kg for 21 days). Dopamine depletion elevates enkephalin, decreases substance P and has little effect on dynorphin. Subsequent D1 agonist treatment has no effect on enkephalin (contained in D2-bearing neurons) but reverses the lesion-induced decrease in substance P and causes a large increase in dynorphin mRNA, both of which are contained in D1-bearing neurons. On the other hand, subsequent to dopamine depletion of the striatum, D2 agonist treatment reverses the lesion-induced elevation of enkephalin mRNA in neurons bearing D2 receptors, but has no effect on substance P or dynorphin in D1 bearing neurons. (Adapted from Gerfen, C. R. et al., *Science*, 250, 1429, 1990.)

This was tested in a study in which selective D1 and D2 agonists were administered to animals with nigrostriatal dopamine lesions.[5] Results demonstrated that continuous treatment with a D2 agonist (quinpirole, 1 mg/kg/day for 21 days) reversed the lesion-induced elevation of enkephalin mRNA levels in indirect striatal neurons without affecting peptide levels in direct projecting neurons. Conversely, intermittent repeated treatment with a D1 agonist (SKF38393, 12.5 mg/kg 1 time per day for 21 days) normalized substance P and greatly elevated dynorphin mRNA levels in direct-projecting neurons without affecting peptide levels in indirect-projecting neurons. These results demonstrated that the segregation of D1 and D2 receptors, respectively, to direct and indirect striatal projection neurons is responsible for the opposite effects of dopamine depletion on peptide gene regulation in these neurons. Of significance is the difference in the treatment regimens, which suggested that D2-mediated gene regulation requires ongoing occupation of the receptor, whereas D1-mediated gene regulation is effective with repeated intermittent receptor activation.

5. DOPAMINE-MEDIATED IMMEDIATE-EARLY GENE REGULATION

While dopamine-mediated gene regulation of striatal peptides in striatal neurons is instructive in terms of the differential effects of D1 and D2 receptor activation, respectively, in the direct and indirect projection neurons, inferences concerning the function of dopamine from such studies are limited. One limitation is that alteration of peptide levels occurs over a relatively long time-course, and sometimes requires repeated drug administration.[5,27] This most likely reflects that induction of genes encoding neuronal peptides is dependent on the induction of transcription factors and thus reflects secondary consequences of receptor-mediated signal transduction mechanisms. Identification of transcription factors that are more directly involved in the early stages of receptor-mediated signal transduction has provided additional tools for gene regulation analysis of neuronal function.[32,33] The most widely used of these is the immediate-early gene c-*fos*.[34]

Immunohistochemical localization of Fos has been used to study dopamine receptor-mediated induction of this immediate early gene.[6,35,36] Studies have demonstrated a similar pattern of changes as observed with changes in peptides in direct and indirect striatal projection neurons. Thus, in the dopamine-depleted striatum, D1 agonist treatment results in the induction of Fos immunoreactivity in direct striatal projection neurons,[6] and, conversely, in the normal animal D2 antagonist treatment results in induction of Fos in indirect striatal projection neurons[6,35] (see also Chapters 2 and 6). One difference between dopaminergic regulation of c-*fos* expression compared with altered peptide mRNA levels is the response time, with Fos immunoreactivity being induced within 1 h of drug administration.

Use of Fos immunoreactivity as a measure of response to dopamine receptor manipulation depends on the extremely low basal level of labeling and subsequent induction of labeling in striatal neurons. Thus, quantitative measure of c-*fos* labeling indicates a threshold response, either a neuron is labeled or not. While such a measure has been qualitatively consistent with D1 agonist induction of Fos in direct striatal projection neurons[6] and D2 antagonist induction in indirect striatal projection neurons,[6,35] results with combined D1 and D2 agonist treatments have been less straightforward. Several studies demonstrate that when D1 and D2 agonist treatments are combined, there is a synergistic response such that there are a greater number of Fos-positive striatal neurons in the dopamine-lesioned striatum after combined D1 and D2 agonist treatment than with D1 agonist treatment alone.[37,38] Such results have been variously interpreted. One interpretation is that the synergistic response involves extrastriatal sites of action of the agonists, since the drug treatments are administered systemically.[37] Another interpretation is that D2 receptor activation is necessary for a full D1 agonist effect to be obtained.[37,38] Such synergistic responses are often described to involve mechanisms in which D1 and D2 receptors are co-expressed in the same neuron population.[39-41]

In a series of studies we have addressed the potentiated response to combined D1 and D2 agonist treatment compared with D1 agonist treatment alone.[7,8] A limitation of many dopamine drug studies is that dopamine receptor agonists or antagonists are administered systemically and changes in the levels of striatal mRNAs are presumed to occur as a consequence of action within the striatum. To address this limitation we infused drugs directly into the striatum to assess their site of action.[7] Results demonstrated that infusion of the D1 agonist SKF 38393 directly into the striatum resulted in the induction of the two immediate-early gene mRNAs encoding Fos and *zif*268 in the infused area. Additionally, the potentiated response to systemic administration of the D1 agonist SKF 38393 and the D2 agonist quinpirole compared to SKF 38393 alone is blocked by intrastriatal infusion of the D2 antagonist eticlopride. These results suggest that the synergistic interaction of D1 and D2 receptor activation occurs within the striatum.

In a second study we used a two-part experiment to address the question of D1-D2 dopamine receptor synergy.[8] In the first experiment we compared the level of c-*fos* and *zif*268 mRNAs in a dose response paradigm and measured the response at the cellular level (Figure 4). Results demonstrated that compared to controls, the number of striatal neurons displaying increased levels of c-*fos* mRNA to a given dose of the D1 agonist, SKF 38393, (0.5-1.5 mg/kg) was increased when combined

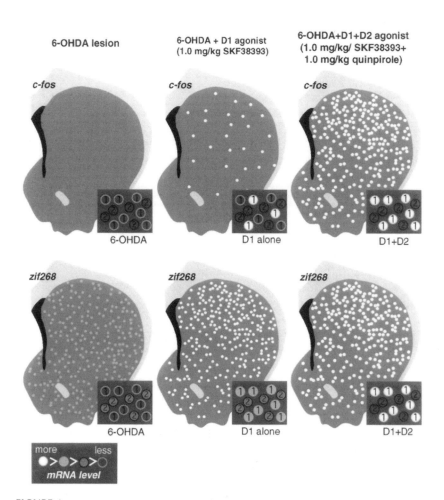

FIGURE 4.
Diagram of immediate-early gene induction by D1 and combined D1 and D2 agonist treatment of rats in dopamine-depleted striatum. The figure shows c-*fos* and *zif*268 mRNA labeling in the dopamine depleted striatum (6-OHDA), in the dopamine-depleted striatum of animals treated with the D1 agonist SKF 38393 (1 mg/kg) alone, and in the dopamine-depleted striatum of animals treated with the D1 agonist SKF 38393 (1 mg/kg) plus the D2 agonist quinpirole (1 mg/kg). (Adapted from Gerfen, C. R. et al., *J. Neuroscience*, 15, 8167, 1995.)

with the D2 agonist, quinpirole, (1 mg/kg). This result is consistent with those obtained with immunohistochemical localization of Fos protein. However, when *zif*268 mRNA levels were measured in these same animals, it was observed that the same number of neurons showed elevated *zif*268 mRNA levels with either D1 agonist alone or when combined with D2 agonist treatment, but the amount of labeling per cell was significantly increased with the combined D1 and D2 agonist treatment. In the latter case, the number of *zif268* labeled striatal neurons showing increased levels to either D1 agonist alone or combined D1 and D2 agonist was roughly half of the striatal neuron population. This result suggested that D1 agonist treatment alone (at doses of 0.5-1.5 mg/kg) resulted in an increase in the expression of the *zif268* in most of the striatal neurons that express the D1 receptor, and that combined D1 and D2 agonist treatment further elevated immediate early gene mRNA levels in this neuron population. This was only inferred based on the number of neurons in which the elevated immediate-early genes was detected.

A direct examination of which striatal neuron populations the altered *zif268* mRNA response occurs in was provided in a second experiment using combined localization of mRNA encoding enkephalin (ENK), which is used as a marker of the D2-containing indirect striatal projection population, and mRNA encoding the immediate-early gene product, *zif268*.[8] In this experiment, enkephalin mRNA was labeled with an immunohistochemical localization of a marker on an antisense probe to enkephalin mRNA, and *zif268* mRNA was localized with a radioactive (^{35}S-UTP) labeled antisense ribonucleotide probe (Figure 5). First, the level of *zif268* mRNA expression was examined in the dopamine-depleted striatum compared with the unlesioned striatum. In the unlesioned striatum, *zif268* mRNA was expressed in both, ENK+ and ENK− neurons, but at a slightly higher level in ENK− neurons. In the dopamine-depleted striatum, ENK+ neurons displayed a significant increase and ENK− neurons displayed a significant decrease in *zif268* mRNA levels, relative to the unlesioned striatum. This result is consistent with the effects of dopamine-depletion on peptides in these neurons.[5,27] Second, the level of *zif268* mRNA was examined in the dopamine-depleted striatum of animals treated with the D1 agonist alone (SKF 38393, 1 mg/kg) compared to the dopamine-depleted striatum of untreated animals. In this case, *zif268* mRNA levels were significantly elevated in ENK− neurons but unchanged in ENK+ neurons. This result is consistent with D1 agonist selectively activating D1 (ENK−) neurons. Notably, the entire population of ENK− neurons showed elevated *zif268* mRNA levels. Finally, the effects of combined D1 (SKF 38393, 1mg/kg) and D2 (quinpirole, 1mg/kg) agonists were compared to that of the D1 agonist alone. In this case, the entire population of ENK− neurons displayed a further significant increase in *zif268* mRNA levels, whereas the entire population of ENK+ neurons displayed a significant decrease in *zif268* mRNA levels. This result suggests that the addition of D2 agonist treatment causes a decrease in the *zif268* mRNA levels in ENK+ D2-containing neurons. The potentiated *zif268* mRNA response in ENK− neurons, putative D1-containing neurons, is thus thought to result from a reduced inhibition by ENK+ neurons of ENK− neurons. The potential of inhibition between indirect

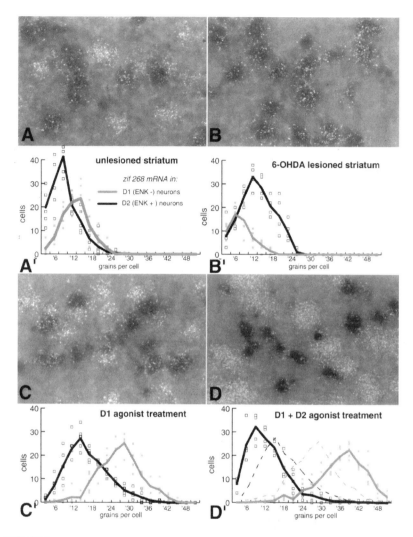

FIGURE 5.
Photomicrographs and frequency distribution of the amount of *zif*268 mRNA labeling in the striatum of animals with 6-OHDA depletion of dopamine in the striatum on one side and treated with D1 and D2 agonists. (A) Unlesioned and vehicle-treated striatum; (B) dopamine-depleted striatum, vehicle treated; (C) dopamine-depleted striatum, D1 agonist (SKF38393, 1 mg/kg) and (D) dopamine-depleted striatum, D1 agonist (SKF 38393, 1 mg/kg) and D2 agonist (quinpirole, 1 mg/kg). All drugs were administered i.p., and the animals killed 60 min following drug injection. Photomicrographs (A,B,C,D) show *zif*268 mRNA labeling (S35-generated silver grains, white) in sections labeled for enkephalin mRNA (black cells labeled with alkaline phosphatase reaction product). Frequency distribution graphs (A′,B′,C′,D′) provide data of the average amount of *zif*268 mRNA label per cell for enkephalin-positive cells, which are putative D2-containing neurons (ENK+, individual cases marked as black squares, average marked as a black line), and enkephalin-negative, putatitive D1-containing cells (ENK–, individual cases marked as gray boxes, average marked as a gray line). (From Gerfen, C. R. et al., *J. Neuroscience*, 15, 8167, 1995.)

(ENK+) and direct (ENK–) neurons is thought to be mediated by the local axon collaterals that provide synaptic contact between these neurons. However, to date there is no direct physiologic evidence of such inhibitory interactions,[42] although the synaptic contacts have been well established.[12]

The following conclusions may be drawn from these studies in the dopamine-depleted striatum. First, D1 agonist treatment alone results in D1 receptor-mediated increased gene expression in the entire population of direct projecting striatal neurons. Second, combined D1 and D2 agonist treatment results in opposite effects on gene regulation in the direct and indirect neuron populations. Third, the potentiated response of direct striatal projection neurons to combined D1 and D2 agonist treatment compared with D1 agonist treatment alone might reflect intercellular interactions rather than interactions between receptors located on the same neurons. A further observation is worth making. While the induction of c-*fos* has been an effective tool for study of dopamine receptor-mediated changes in striatal neurons it is limited in two ways. First, it provides information concerning increased levels of activation of neurons, but not decreases. Here the immediate early gene *zif*268 is useful, because its constitutive expression allows not only increases but also decreases in expression to be observed. Second, the induction of c-*fos* appears to reflect a threshold effect but does not provide quantitative measure of graded responses. Again, *zif*268 appears to provide a useful marker for such responses. These limitations of the use of c-*fos* as a marker of gene regulation should not be taken to suggest that it does not have utility, but rather that conclusions drawn from observation of its expression in gene regulation paradigms are augmented when combined with other markers.

6. STRIATAL ADAPTIVE RESPONSES: DOPAMINE-OPIOID INTERACTIONS

Most of the studies discussed thus far have employed the model in which dopamine is depleted in the striatum. We have used this model primarily because it allows examination of D1 and D2 receptor-mediated effects in isolation of one another to address the question of the striatal neuron population affected by manipulation of each receptor subtype. However, several limitations are inherent in this model. First, the response of striatal neurons in the dopamine-depleted state is supersensitive, particularly to D1 receptor stimulation. This raises questions as to whether the conclusions drawn from such studies are applicable to the normal function of dopamine in the striatum. While this is a reasonable criticism it should be stated that certain conclusions drawn from such studies are valid. Most notably it does not appear that in the dopamine-depleted striatum the phenotypic characteristics of striatal neurons (in terms of their expression of dopamine receptor subtypes or neuropeptides) are altered.[5] Nonetheless, the supersensitivity of the D1 responsiveness of the direct projection neurons needs to be taken into consideration. Second, the normal striatum displays considerable heterogeneity such that the relative expression of various neuronal markers and the relative response of these neurons to dopamine receptor manipulation vary both regionally and in the patch and matrix compartments.[3,11]

In order to examine regional variations in dopamine receptor-mediated gene regulation in the striatum we have employed the indirect dopamine agonist cocaine. The rationale behind using this paradigm is that cocaine, which prolongs the efficacy of released dopamine by blocking its re-uptake by the dopamine transporter, elicits the induction of c-*fos* in the striatum of normal rats.[9,10,43-45] This enables examination of dopamine receptor-mediated gene regulation in which the receptor response had not been rendered supersensitive. In the first study, following a dose response experiment, cocaine was administered at a dose of 30 mg/kg twice a day for 4 days, and the levels of c-*fos* and peptide mRNAs analyzed in the striatum on each of those days at a timepoint 30 min after the drug administration.[9] Results demonstrated the following (Figure 6). After the first injection of cocaine there was a significant induction of c-*fos* mRNA in the dorsal striatum. A similar pattern and level of induction was also observed with a second day treatment, but the level of induction decreased after the third and fourth day injections. We also observed that the regional pattern of cocaine-induced c-*fos* expression after the first day injection was complementary to the pattern of dynorphin mRNA expression in the striatum. That is, dynorphin mRNA, which is expressed in direct projecting neurons throughout the striatum, has a higher basal level of expression on a per cell basis in the ventral striatum, including the nucleus accumbens, than in the dorsal striatal region in which cocaine-induced c-*fos* mRNA levels were elevated. None of the other peptide mRNA levels measured, including enkephalin and substance P, displayed this complementary pattern. With successive cocaine injections the levels of dynorphin mRNA were elevated in the dorsal striatal region, coincident with the decreasing response of c-*fos* mRNA to these injections. This led us to propose that dynorphin functions to "blunt" the cocaine induction of c-*fos* gene expression and that repeated cocaine administration results in an adaptive response consisting of increases in dynorphin mRNA expression. To test this hypothesis, spiradoline, an agonist of the kappa opioid receptor, which is the receptor dynorphin is thought to act on, was co-administered with cocaine systemically in

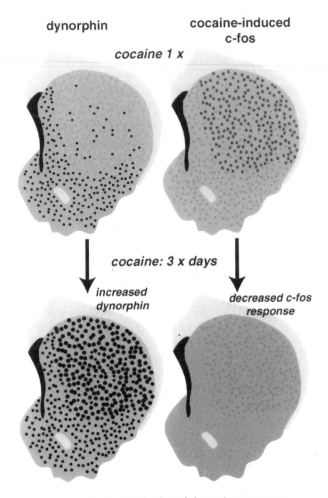

regional organization of dopamine response

FIGURE 6.

Diagram of the regional response within the striatum to the indirect dopamine agonist cocaine demonstrating the functional role of dynorphin in modulating this response. The basal level of dynorphin expression shows a higher level in ventral and medial striatal regions. A single injection of cocaine induces the immediate-early gene c-*fos* by a D1-mediated mechanism in the dorsolateral striatal region, complementary to the area showing high levels of dynorphin. Repeated treatment with cocaine (twice daily injections of 30 mg/kg for 3–4 days) results in an increase in dynorphin levels in the dorsal striatal region, which has low basal expression, and a marked reduction of c-*fos* induction in this area, in which c-*fos* had previously been induced. These data suggest that dynorphin blunts the response of neurons to D1 receptor stimulation. Further studies have shown that this effect of dynorphin is mediated through kappa opioid receptors. (Adapted from Steiner, H. and Gerfen, C. R., *J. Neuroscience*, 13, 5066, 1993.)

drug naive animals. In this case, cocaine-induced c-*fos* mRNA expression in the dorsal striatum was markedly reduced. This result substantiated our proposed action of dynorphin in blunting the responses of striatal projection neurons to excessive dopamine receptor activation. In the second study two points regarding the role of dynorphin in the response to cocaine were clarified.[10] The first was the characterization of the dopamine receptor subtype mediating immediate-early gene induction in response to cocaine. When the D1 antagonist SCH23390 was infused directly into the striatum prior to a systemic administration of cocaine, the induction of c-*fos* and *zif* 268 mRNA was blocked. This demonstrated that the cocaine induction of immediate-early genes in the dorsal striatum is a D1 receptor-mediated response. This is consistent with the observation that cocaine treatment results in elevated dynorphin and substance P mRNA levels in neurons which express the D1 dopamine receptor.[44] Secondly, though our previous study demonstrated that the kappa opioid receptor agonist spiradoline blocks cocaine-induced c-*fos* mRNA in the dorsal striatum, the systemic administration of the drug in that study limited our ability to conclude whether or not the action of the dynorphin agonist occurred within the striatum. To examine this question directly, spiradoline was infused directly into the striatum. When administered in this fashion also, spiradoline blocked cocaine-induced immediate early gene expression in the dorsal striatum.

The results of these studies suggest a functional role for dynorphin in the striatum to modify the response of direct striatal projection neurons to excessive D1 receptor stimulation. Evidence in support of such a function comes first from the finding that the D1-mediated cocaine induction of immediate early genes in the striatum occurs in the dorsal striatal region in which dynorphin is normally at a relatively low level, and it does not occur in regions with constitutively high levels of dynorphin. Secondly, repeated cocaine treatment results in an adaptive response in the dorsal region that involves induction of dynorphin levels coincident with a decreased D1-mediated cocaine induction of immediate early genes. Thirdly, pharmacologic treatment with a dynorphin agonist is able to mimic the adaptive response of elevated dynorphin levels by blocking D1-mediated immediate early gene induction in the dorsal striatum.

Among the questions that remain concerning the action of dynorphin is the mechanism by which it affects the D1 receptor-mediated response. Dynorphin is thought to act primarily at kappa opioid receptor sites. *In situ* hybridization histochemical studies have identified kappa receptor mRNA at very low levels in striatal neurons themselves, and then highest levels occur in the ventral striatum.[46] The most abundant level of expression appears to be in the dopamine neurons in the substantia nigra pars compacta, which provide the dopamine input to the striatum. Thus, one possibility is that dynorphin acts presynaptically to limit the release of dopamine in the striatum. To address this possibility we conducted a study using the D1 agonist induction of immediate early genes in the dopamine-depleted striatum.[47] In this case, systemic administration of the kappa receptor agonist, spiradoline, was ineffective at decreasing the response to D1-agonist inducted of striatal immediate early gene expression. Moreover, repeated daily injections of the D1 agonist SKF 38393 (2 mg/kg) for 4 days did not reduce the level of induction of immediate-early genes, although the level of dynorphin mRNA resulting from such treatments was elevated considerably. Thus, the adaptive response to excessive D1 receptor stimulation that results in elevated dynorphin expression in the dopamine-depleted striatum does not produce the same attenuation of response as occurs to cocaine treatment in the normal striatum. There are several explanations for this difference. One is that in the normal striatum with an intact dopamine innervation, dynorphin acts presynpatically to reduce dopamine release and thus blunt the response of D1 receptor stimulation. While this possibility is consistent with the localization of kappa receptors on dopamine terminals, it is also possible that in the dopamine-depleted striatum the supersensitive D1 receptor response that results in the induction of immediate early genes is unaffected by dynorphin. What is apparent is that in the dopamine-depleted striatum the mechanisms that are normally in place to provide for adaptive attenuation of neuronal response to repeated D1 receptor stimulation are ineffective.

7. CONCLUSIONS

The purpose of the discussed studies has been to elucidate the functional role of dopamine in the striatum. To accomplish this, studies were designed to examine D1 and D2 receptor-mediated alterations in gene regulation in defined striatal neurons. The intent of each study was to isolate, as well as possible, the receptor-mediated effects and to define the neurons in which such effects occur. This approach has several limitations, among which is the uncertainty of the receptor-mediated signal transduction mechanisms that lead to the measured alterations of mRNA expression (see Chapter 5). Another limitation of this approach is the fact that alterations in gene regulation are not correlated with changes in the physiologic response of these neurons. Nonetheless, a number of observations have been obtained that have extended our understanding of the functional role of dopamine in the striatum.

First, *in situ* hybridization histochemical localization of mRNAs encoding peptides and dopamine receptor subtypes allowed for the characterization of the direct striatal projection neurons as expressing the D1 dopamine receptor and dynorphin and substance P, and the indirect striatal projection neurons as expressing the D2 dopamine receptor and enkephalin.[5,28,29] The dopamine-depleted striatum has provided a model in which dopamine receptor-mediated gene regulation demonstrates that the direct effects of D1 and D2 receptor agonists alter genes, respectively, in the direct and indirect striatal neuron populations that express these receptors. Evidence comes from the findings that elevated enkephalin mRNA levels in the indirect striatal neurons resulting from dopamine depletion are reversed selectively by D2 agonists, whereas decreased levels of the peptides in the direct striatal neurons are elevated selectively by D1 agonist treatment.[5] These opposite effects of dopamine agonists on these two neuron populations were confirmed with studies demonstrating simultaneous decreases and increases in immediate early genes, respectively, in indirect and direct striatal neurons when D2 and D1 agonists are used together.[8] The fact that such effects result in a synergistic response further suggests that interactions between D1 and D2 dopamine receptor subtypes occur through interactions between neurons, each expressing only one of the receptor subtypes, rather than by interactions of the effects of these receptors on single neurons.

Second, the cocaine model provided insight into the interactions between dopamine and peptide functions in the striatum.[9,10] Studies utilizing the dopamine-depleted striatal model are limited by the fact that the response of direct D1-containing striatal neurons is supersensitive to D1 receptor stimulation. The cocaine model is useful because it provides insight into the "normal" striatum. What is apparent in the normal striatum is the regional heterogeneity of peptide levels. For example, dynorphin is expressed by direct striatal neurons throughout the striatum, but at higher levels per cell in the ventral striatum. The finding that cocaine results in the induction of immediate early genes in the dorsal region of the striatum, which is complementary to the area in which dynorphin is expressed at high levels, suggested a correlation between dynorphin expression and the cocaine response. The subsequent elevation of dynorphin and reduction of the cocaine-induced immediate early gene response that results from repeated cocaine treatment suggests further that the expression of dynorphin is an adaptive response to excessive D1 receptor stimulation that functions to blunt such stimulation. These findings not only suggest that in the cocaine paradigm there is an adaptive response of striatal neurons to produce dynorphin, but they also suggest that in the normal striatum the regional heterogeneity of neuropeptide expression is related to the normal differences of dopamine neurotransmission in different regions of the striatum. Such differences may reflect differences in dopamine systems directed to the ventral striatum compared to those directed to the dorsal striatum, or to differences in the pattern of activity of cortical inputs to these different striatal regions.

REFERENCES

1. Gerfen, C.R., Herkenham, M. and Thibault, J., The neostriatal mosaic: II. Patch and matrix directed mesostriatal dopaminergic and non-dopaminergic systems, *J. Neurosci.*, 7, 3915, 1987.
2. Albin, R.L., Young, A.B. and Penney, J.B. The functional anatomy of basal ganglia disorders, *Trends Neurosci.*, 12, 366, 1989.
3. Gerfen, C.R., The neostriatal mosaic: Multiple levels of compartmental organization, *Trends Neurosci.*, 15,133, 1992.
4. Kebabian, J.W. and Calne, D.B., Multiple receptors for dopamine, *Nature*, 277, 93, 1979.
5. Gerfen, C.R., Engber, T.M., Mahan, L.C., Susel, Z., Chase, T.N., Monsma, F.J., Jr. and Sibley, D.R., D1 and D2 dopamine receptor-regulated gene expression of striatonigral and striatopallidal neurons, *Science*, 250,1429, 1990.
6. Robertson, G.S., Vincent, S.R. and Fibiger, H.C., D1 and D2 dopamine receptors differentially regulate c-*fos* expression in striatonigral and striatopallidal neurons, *Neuroscience*, 49, 285, 1992.
7. Keefe, K. A. and Gerfen, C.R., D1-D2 Dopamine receptor synergy in striatum: effects of intrastriatal infusions of dopamine agonists and antagonists on immediate early gene expression, *Neuroscience*, 66, 903, 1995.
8. Gerfen, C.R., Keefe, K.A. and Gauda, E.B., D1 and D2 dopamine receptor function in the striatum: Coactivation of D1- and D2-dopamine receptors on separate populations of neurons results in potentiated immediate early gene response in D1-containing neurons, *J. Neurosci.*, 15, 8167, 1995.
9. Steiner, H. and Gerfen, C.R., Cocaine-induced c-*fos* messenger RNA is inversely related to dynorphin expression in striatum, *J. Neurosci.*, 13,5066, 1993.
10. Steiner, H. and Gerfen, C.R., Dynorphin opioid inhibition of cocaine-induced, D1 dopamine receptor-mediated immediate early gene expression in the striatum, *J. Comp. Neurol.*, 353, 200, 1994.
11. Gerfen, C.R. and Wilson, C.J., The basal ganglia. In: *Handbook of Chemical Neuroanatomy*, Hokfelt, T., Bjorklund, A., and Swanson, L.W., eds. Elsevier, 1996, 381.
12. Wilson, C.J. and Groves, P.M., Fine structure and synaptic connections of the common spiny neuron of the rat neostriatum: a study employing intracellular inject of horseradish peroxidase, *J. Comp. Neurol.*, 194, 599, 1980.
13. Kawaguchi, Y., Wilson, C.J. and Emson, P.C., Projection subtypes of rat neostriatal matrix cells revealed by intracellular injection of biocytin, *J. Neurosci.*, 10, 3421, 1990.
14. Deniau, J.M. and Chevalier, G., Disinhibition as a basic process in the expression of striatal functions. II. the striato-nigral influence on thalamocortical cells of the ventromedial thalamic nucleus, *Brain Res.*, 334, 227, 1985.
15. Kita, H., Chang, H.T. and Kitai, S.T., Pallidal inputs to subthalamus: intracellular analysis, *Brain Res.*, 264, 255, 1983.

16. Kita, H. and Kitai, S.T., Efferent projections of the subthalamic nucleus in the rat: light and electron microscopic analysis with the PHA-L method, *J. Comp. Neurol.*, 260, 435, 1987.

17. Bergman, H., Wichmann, T. and DeLong, M.R., Reversal of experimental parkinsonism by lesions of the subthalamic nucleus, *Science*, 249, 1436, 1990.

18. Gerfen, C.R. and Young, W.S., Distribution of striatonigral and striatopallidal peptidergic neurons in both patch and matrix compartments: an *in situ* hybridization histochemistry and fluorescent retrograde tracing study, *Brain Res.*, 460, 161, 1988.

19. Kita, H. and Kitai, S.T., Glutamate decarboxylase immunoreactive neurons in rat neostriatum: their morphological types and populations, *Brain Res.*, 447, 346, 1988.

20. Le Moine, C. and Bloch, B., D1 and D2 dopamine receptor gene expression in the rat striatum: sensitive cRNA probes demonstrate prominent segregation of D1 and D2 mRNAs in distinct neuronal populations of the dorsal and ventral striatum, *J. Comp. Neurol.*, 355, 418, 1995.

21. Hong, J.S., Yang, H.-Y.T., Fratta, W. and Costa, E., Rat striatal methionine-enkephalin content after chronic treatment with cataleptogenic and noncataleptogenic drugs, *J. Pharmacol. Exp. Ther.*, 205, 141, 1978.

22. Hanson, G.R., Merchant, K.M., Letter, A.A., Bush, L. and Gibb, J.W., Methamphetamine-induced changes in the striato-nigral dynorphin sytem: role of D-1 and D-2 receptors, *Eur. J. Pharmacol.*, 144, 245, 1987.

23. Young, W.S. III, Bonner, T.I. and Brann, M.R., Mesencephalic dopaminergic neurons regulate the expression of neuropeptide mRNAs in the rat forebrain, *Proc. Natl. Acad. Sci. U.S.A.*, 83, 9827, 1986.

24. Hong, J.S., Yang, H.-Y.T. and Costa, E., Projections of substance P containing neurons from neostriatum to substantia nigra, *Brain Res.*, 121, 541, 1977.

25. Haber, S.N. and Watson, S.J., The comparison between enkephalin-like and dynorphin-like immunoreactivity in both monkey and human globus pallidus and substantia nigra, *Life Sci.*, 1, 33, 1983.

26. Graybiel, A.M. and Chesselet, M.F., Compartmental distribution of striatal cell bodies expressing [Met]enkephalin-like immunoreactivity, *Proc. Natl. Acad. Sci. U.S.A.*, 81, 7980, 1984.

27. Gerfen, C.R., McGinty, J.F. and Young, W.S., Dopamine differentially regulates dynorphin, substance P, and enkephalin expression in striatal neurons: *In situ* hybridization histochemical analysis, *J. Neurosci.*, 11, 1016, 1991.

28. Le Moine, C., Normand, E., Guitteny, A.F., Fouque, B., Teoule, R. and Bloch, B., Dopamine receptor gene expression by enkephalin neurons in rat forebrain, *Proc. Natl. Acad. Sci. U.S.A.*, 87, 230, 1990.

29. Le Moine, C., Normand, E. and Bloch, B., Phenotypical characterization of the rat striatal neurons expressing the D1 dopamine receptor gene, *Proc. Natl. Acad. Sci. U.S.A.*, 88, 4205, 1991.

30. Levey, A.I., Hersch, S.M., Rye, D.B., Sunahara, R.K., Niznik, H.B., Kitt, C.A., Price, D.L., Maggio, R. and Brann, M.R., Localization of D1 and D2 dopamine receptors in brain with subtype-specific antibodies, *Proc. Natl. Acad. Sci. U.S.A.*, 90, 8861, 1993.

31. Hersch, S.M., Ciliax, B.J., Gutekunst, C.-Y., Rees, H.D., Heilman, C.J., Uung, K.K.L., Bolam, J.P., Ince, E., Yi, H. and Levey, A.I., Electron microscopic analysis of D1 and D2 dopamine receptor proteins in the dorsal striatum and their synaptic relationships with motor corticostriatal afferents, *J. Neurosci.*, 15, 5222, 1995.

32. Curran, T., Gordon, M.B., Rubino, K.L. and Sambucetti, L.C., Isolation and characterization of the c-*fos*(rat) cDNA and analysis of post-translational modification *in vitro*, *Oncogene*, 2, 79, 1987.

33. Milbrandt, J., A nerve growth factor-induced gene encodes a possible transcriptional regulatory factor, *Science*, 238, 797, 1987.

34. Morgan, J.I. and Curran, T., Stimulus-transcription coupling in neurons: role of cellular immediate-early genes, *Trends Neurosci.*, 12, 459, 1989.

35. Dragunow, M., Robertson, G.S., Faull, R.L.M., Robertson, H.A. and Jansen, K., D2 dopamine receptor antagonists induce Fos and related proteins in rat striatal neurons, *Neuroscience*, 37, 287, 1990.

36. Robertson, G.S., Herrera, D.G., Dragunow, M. and Robertson, H.A., L-dopa activates c-*fos* in the striatum ipsilateral to a 6-hydroxydopamine lesion of the substantia nigra, *Eur. J. Pharmacol.*, 159, 99, 1989.

37. Paul, M.L., Graybiel, A.M., David, J.-C. and Robertson H.A., D1- and D2-like dopamine receptors synergistically activate rotation and c-*fos* expression in the dopamine-depleted striatum in a rat model of Parkinson's disease, *J. Neurosci.*, 12, 3729, 1992.

38. LaHoste, G.J., Yu, J. and Marshall, J.F., Striatal Fos expression is indicative of dopamine D1/D2 synergism and receptor supersensitivity, *Proc. Natl. Acad. Sci. U.S.A.*, 90, 7451, 1993.

39. Bertorello, A.M., Hopfield, J.F., Aperia, A. and Greengard, P., Inhibition by dopamine of $(Na^{++}K^+)$ATPase activity in neostriatal neurons through D1 and D2 dopamine receptor synergism, *Nature*, 347, 386, 1990.

40. Piomelli, D., Pilon, C., Giros, B., Sokoloff, P., Martres, M.-P. and Schwartz, J.-C., Dopamine activation of the arachidonic acid cascade as a basis for D1/D2 receptor synergism, *Nature*, 353, 164, 1991.

41. Surmeier, D.J., Eberwine, J., Wilson, C.J., Cao, Y., Stefani, A. and Kitai, S.T., Dopamine receptor subtypes colocalize in rat striatonigral neurons, *Proc. Natl. Acad. Sci. U.S.A.*, 89, 10178, 1992.

42. Jaeger, D., Kita, H. and Wilson, C.J., Surround inhibition among projection neurons is weak or nonexistent in the rat neostriatum, *J. Neurophysiol.*, 72, 2555, 1994.

43. Graybiel, A.M., Moratalla, R. and Robertson, H.A., Amphetamine and cocaine induce drug-specific activation of the c-*fos* gene in striosome-matrix compartments and limbic subdivisions of the striatum, *Proc. Natl. Acad. Sci. U.S.A.*, 87, 6912, 1990.

44. Cenci, M.A., Campbell, K., Wictorin, K. and Björklund, A., Striatal c-*fos* induction by cocaine or apomorphine occurs preferentially in output neurons projecting to the substantia nigra in the rat, *Eur. J. Neurosci.*, 4, 376, 1992.

45. Young, S.T., Porrino, L.J. and Iadarola, M.J., Cocaine induces striatal c-Fos-immunoreactive proteins via dopaminergic D_1 receptors. *Proc. Natl. Acad. Sci. U.S.A.*, 88, 1291, 1991.

46. Mansour, A., Fox, C.A., Meng, F., Akil, H. and Watson, S.J., k_1 receptor mRNA distribution in the rat CNS: comparison to k receptor binding and prodynorphin mRNA, *Mol. Cell. Neurosci.*, 5, 124, 1994.

47. Steiner, H. and Gerfen, C.R., Dynorphin regulates D1 dopamine receptor-mediated responses in the striatum: relative contributions of pre- and postsynaptic mechanisms in dorsal and ventral striatum demonstrated by altered immediate-early gene induction, *J. Comp. Neurol.*, in press.

Chapter **2**

ACUTE AND LONG-TERM ALTERATIONS IN IMMEDIATE EARLY GENE EXPRESSION BY ANTIPSYCHOTIC DRUGS: DOWNSTREAM GENOMIC TARGETS, BEHAVIORAL CORRELATES, AND ROLE OF D2, D3, AND D4 RECEPTORS

Kalpana M. Merchant

CONTENTS

0-8493-8550-4/96/$0.00+$.50
© 1996 by CRC Press, Inc.

1. INTRODUCTION

Antipsychotic drugs (APDs) are a group of chemically diverse agents used for symptomatic treatment of psychotic disorders like schizophrenia. Clinically used APDs display a variety of pharmacological effects including blockade of dopamine, serotonin, muscarinic, histamine and noradrenaline receptors. However, blockade of dopamine D2-like receptors (see Section 4.1 for classification of dopamine receptors) appears to be crucial for the effects of APDs since the therapeutic potency of these drugs correlates well with their affinity for dopamine D2-like receptors.[1-3] In contrast, there is poor correlation between clinical efficacy and binding to serotonin 5HT2, muscarinic, histamine-H1 and noradrenergic α1 blockade.[3] This observation coupled with psychotogenic effects of indirect dopamine agonists like amphetamine forms the major basis for the "dopamine hypothesis" of schizophrenia.

Clinical studies of schizophrenic patients indicate that the currently available APDs differ markedly in their efficacy against positive (e.g., hallucinations, delusions) vs. negative (e.g., flattening of affect, speech, anhedonia) symptoms and side effect (motor and endocrine) profile. This has led to the classification of APDs into the so-called typical APDs, or neuroleptics, and atypical APDs.[4,5] Atypical agents, typified by clozapine, produce little or no extrapyramidal motor side effects (EPS) like parkinsonism and tardive dyskinesia, do not produce hyperprolactinemia, show superior efficacy against negative symptomatology and are often effective in treating neuroleptic-resistant patients.[6] However, pharmacological mechanisms subserving the superior therapeutic effects of clozapine are unclear.[7,8] Clinical use of clozapine is associated with serious side effects such as agranulocytosis and seizures. Hence, studies to gain an insight into the mechanisms underlying the superior antipsychotic efficacy of clozapine is an area of highly pursued research that may lead to better APDs.

One of the strategies being followed by a number of laboratories is to identify and trace neurochemical pathways targeted by typical vs. atypical APDs in order to determine whether anatomical specificity contributes to the differences in the clinical profiles of the two classes of APDs. This approach is based on the vast literature of neuroimaging, behavioral and physiological studies that indicate that discrete neural systems contribute to the specific symptoms of schizophrenia.[9,10] For example, prefrontal cortical systems are thought to play a role in manifestation of negative symptoms and cognitive deficits while the positive symptoms may be related to alterations in neural activity in subcortical limbic structures such as the nucleus accumbens and the hippocampal formation. On the other hand, extrapyramidal motor side effects of neuroleptics may be related primarily to changes in the caudatoputamen.

APDs produce their full therapeutic effects after a few weeks of treatment. This suggests that long-term phenotypic changes in targeted neurons may underlie the therapeutic efficacy of these drugs. Hence, studies aimed at identification of neuroanatomic targets of APDs have taken advantage of the ability of immediate-early genes (IEG) as reporters of alterations in neuronal activity.[11,12] APD-induced alterations in the expression of IEGs encoding transcription regulatory factors such as Fos, FosB and *Zif*268 (also known as NGFI-A, Egr-1, Krox-24) [13-15] have been examined by Northern and Western blot analysis as well as *in situ* hybridization histochemistry and immunohistochemistry.[17-22]

In this chapter I will review first the data on regional alterations in the expression of IEGs by acute administration of APDs and possible downstream targets of these transcription factors. Next, the effects of chronic APD treatment on regional gene expression and possible correlation with behavioral effects of the APDs will be examined. Finally, the involvement of newly identified subtypes of dopamine D2-like receptors in the region-specific IEG induction will be reviewed.

2. ACUTE EFFECTS OF APDs ON IEG EXPRESSION

2.1 Induction of c-*fos* Gene Expression

Fos, the protein product encoded by the IEG, c-*fos*, is one of the AP-1 transcription factors.[13] Fos forms heterodimers with other members of the AP-1 transcription factors and regulates transcription of targeted genes via AP-1 recognition sequences. Changes in c-*fos* mRNA and Fos protein levels are well-characterized phenotypic effects of APDs. Acute alterations in c-*fos* gene expression by APDs are thought to signal alterations in neuronal activity that lead to long-term phenotypic changes in cells targeted by the APDs.

2.1.1 Alterations in c-*fos* Gene Expression in the Caudate-Putamen

Miller[16] and Dragunow et al.[17] first demonstrated that acute administration of the neuroleptic, haloperidol, increases c-*fos* expression in the neostriatum. This observation led to studies comparing the induction of c-*fos* in the striatum by haloperidol with that produced by clozapine, the prototype of atypical APDs.[18-20] The results demonstrated that whereas haloperidol robustly induced c-*fos* in the dorsolateral sector of the striatum (DLSt), clozapine had no effect on Fos expression in this region. The DLSt is primarily a "motor" region of the caudatoputamen.[21,22] Hence, the induction of c-*fos* by haloperidol but not clozapine suggests that the DLSt Fos response may be an index of the high EPS liability of haloperidol. To investigate this possibility, we characterized the effects of five clinically tested typical and atypical APDs from different chemical classes.[23] We ensured that the various APDs were compared at doses producing equivalent biological effects in rats. Thus the effects of the APDs on c-*fos* mRNA induction were tested at doses 1 to 10 times their ED_{50} for blockade of apomorphine-induced hyperlocomotion in rats.[24] The butyrophenone, haloperidol, and the phenothiazine, fluphenazine, two APDs with high EPS liability, increased the expression of c-*fos* mRNA in the DLSt at doses 1 to 2 times their ED_{50} for blockade of apomorphine-induced locomotor activation.

On the other hand, clozapine, remoxipride and thioridazine did not induce Fos significantly in the DLSt even at doses up to 10 times their ED_{50} for blockade of hyperlocomotion produced by apomorphine. As discussed above, clozapine is the prototype of atypical agent.[6] Remoxipride was tested in the clinic and found to have an atypical profile as compared to both haloperidol and fluphenazine.[25-27] However, its use has been halted due to a risk of aplastic anemia.[28] Thioridazine is an APD with an apparently atypical profile in rodents.[24,29] In human patients it produces relatively fewer motor side effects as compared to neuroleptics.[30] These data are consistent with the idea that c-*fos* induction in the DLSt neurons may be an index of EPS liability of the APD. This view got strong support from a recent report by Robertson et al.[31] who compared the induction of Fos-like immunoreactivity by 17 APDs displaying a range of clinically tested EPS liability. The results showed that Fos induction in the DLSt predicted the EPS liability of an APD reasonably well. Several recent studies have further confirmed the induction of c-*fos* in the DLSt by neuroleptic drugs.[32-34]

It is important to note that even the atypical agents, including clozapine, do induce c-*fos* in the DLSt, albeit at very high doses.[23,35] There appears to be a positive correlation between the atypical profile of an APD (with respect to lack of EPS) and the dose range that separates its blockade of apomorphine-induced hyperactivity and induction of c-*fos* in the DLSt. For example, clozapine blocks apomorphine-induced locomotor activity at ED_{50} of 4 mg/kg. However, an induction of c-*fos* mRNA is not seen until 60 mg/kg.[23,35] On the other hand, haloperidol begins to induce c-*fos* at 0.1 mg/kg, the same dose at which it blocks apomorphine-induced locomotor activity (unpublished data).

Induction of c-*fos* in the DLSt neurons is selective for the D2-like receptors. The D1-like antagonist, SCH 23390, does not alter c-*fos* expression in the caudatoputamen, nor does it alter the effects of haloperidol.[18,20,23] However, muscarinic[36] and glutamatergic[18] receptors appear to reduce haloperidol-induced c-*fos* expression at relatively high doses indicating converging influence of other neurotransmitter systems on D2-mediated effects (see Chapter 4 for a discussion on this issue).

2.1.1.1 *Downstream Genomic Targets of Fos in the DLSt*

Using retrograde labeling and co-expression of proenkephalin mRNA, Robertson et al. determined that haloperidol-induced Fos expression is predominantly in the medium spiny striatal neurons projecting to the globus pallidus[37] (for a detailed description of the organization of the striatal circuitry, see Chapters 1 and 6). These data are consistent with the findings of Gerfen et al. (Chapter 1) that D2-like receptors regulate the activity of striatopallidal neurons. These neurons constitutively express the opioid peptide, enkephalin, and the inhibitory neurotransmitter GABA.[38,39] An increase in the output of the striatopallidal neurons is thought to produce hypokinetic effects such as catalepsy in rodents and parkinsonism in human patients.[40,41] Increased activity of the striatopallidal neurons is likely to enhance the biosynthesis of the neurotransmitters utilized by these neurons. Hence, it raises the question: Does neuroleptic-induced Fos directly target genes encoding proenkephalin or GABA synthetic enzyme, glutamic acid decarboxylase (GAD)? The proenkephalin gene promoter has AP-1 consensus sequences that regulate the expression of this gene.[42] However, haloperidol increases striatal proenkephalin and GAD mRNA levels only after chronic treatment[43-47] (see also Chapter 7). Hence, neither proenkephalin nor GAD genes are likely to be the direct genomic targets of Fos induced by acute neuroleptic treatment. This is consistent with the observations of Konradi et al.,[48] who demonstrated that in the striatum, haloperidol-induced proenkephalin mRNA expression is regulated by cAMP-response element binding protein, not Fos.

Expression of the gene encoding another neuropeptide, neurotensin, is robustly enhanced by neuroleptic drugs, specifically in the DLSt.[49-50] The increases in neurotensin/neuromedin N (NT/N) mRNA levels were preceded by an increase in the expression of primary transcripts of the NT/N gene, indicating an increase in the transcription of the gene.[51] Thus after a single injection of haloperidol the following temporal pattern of cellular events was observed: (1) an induction of c-fos mRNA, (2) increases in NT/N primary transcripts, (3) increases in the levels of mature NT/N mRNA and (4) increases in neurotensin immunoreactive content.[38,52-55] These data along with the observation that the NT/N gene expression is regulated by AP-1 factors[56] led us to hypothesize that the induction of Fos produced by acute haloperidol administration may increase NT/N gene transcription. A confirmation of this hypothesis was obtained first in studies that showed that following haloperidol, a majority of neurons expressing c-fos mRNA or Fos-like immunoreactivity co-express NT/N mRNA.[57,58] Secondly, intrastriatal administration of an antisense c-fos oligo attenuated haloperidol-induced NT/N mRNA expression.[58,59] Thus the NT/N gene is at least one of the targets of Fos induced by APDs in the DLSt neurons.

2.1.2   Alterations in c-fos and NT/N Gene Expression by APDs in the Nucleus Accumbens

Unlike the DLSt, acute administration of typical as well as atypical APDs induced Fos-like immunoreactivity in the nucleus accumbens.[19,20,31-34] The shell and the core divisions of the nucleus accumbens displayed some specificity for the effects of typical vs. atypical APDs.

Thus only neuroleptics increased Fos immunoreactivity in the core. However, the accumbal shell showed an induction of Fos after acute administration of all clinically effective APDs tested. Interestingly, all APDs also increased NT/N mRNA expression in the accumbal shell.[22,35] Although c-*fos* mRNA levels were enhanced in the most rostral pole of the nucleus accumbens, there was no significant induction of c-*fos* mRNA at the level of the accumbens that showed NT/N gene expression.[22] The double labeling study also failed to show co-expression of neurotensin and c-*fos* mRNA in the accumbal neurons.[57] The reasons underlying the possible discrepancy between Fos protein and c-*fos* mRNA remain unclear. One possibility is that in the nucleus accumbens, APD-induced Fos expression may reflect post-transcriptional alterations. Another possibility that needs investigation is that the time course of c-*fos* mRNA in the accumbens may be significantly more rapid than that in the DLSt. Effects of APDs on c-*fos* and NT/N mRNA were examined at 1 h after treatment[22] by which time accumbal c-*fos* mRNA levels may have declined significantly.

It appears that the neurons of the accumbal shell are a common target of clinically effective antipsychotic drugs. Unlike the core of the nucleus accumbens that resembles the dorsally situated motor region of the caudatoputamen, the shell section of the nucleus accumbens has predominantly limbic characteristics with respect to afferent and efferent connections as well as biochemical alterations.[60,61] Additionally, hyperactivity of dopaminergic systems in the nucleus accumbens is thought to underlie psychotic symptoms.[11,12] Thus it appears that changes in the activity of the accumbal shell may contribute to the one common clinical effect of all APDs tested, i.e., efficacy in the treatment of positive symptoms. The induction of NT/N gene by all APDs raises the question whether alterations in the activity of the neurotensin neurons may contribute to the clinical efficacy of APDs. This possibility is supported by several studies that demonstrate that the pharmacological effects of this peptide resemble those of neuroleptics.[62] Particularly, intra-accumbens administration of neurotensin antagonizes the locomotor activation produced by amphetamine, a behavioral test that appears to correlate with antipsychotic efficacy.[63,64] Further evidence of the involvement of neurotensin in the pathophysiology and pharmacotherapy of schizophrenia is evident in (1) a reduction in neurotensin levels in the CSF of a subgroup of schizophrenic patients which normalizes upon treatment with neuroleptic drugs[65] and (2) increases in neurotensin receptor binding sites in the entorhinal cortex in postmortem schizophrenic tissue.[66]

A recent report demonstrates co-expression of proenkephalin mRNA and Fos-immunoreactivity in the accumbal shell after acute clozapine treatment.[67] Therefore, this opioid peptide may reflect another possible target of APDs in the accumbal shell.

2.1.3 Induction of c-fos Gene Expression in the Medial Prefrontal Cortex

Imaging studies of schizophrenic patients indicate that the activity of the prefrontal cortex is altered in these patients.[68] Anatomical, behavioral and physiological studies of primates as well as rodents and imaging studies in humans indicate that this region may be involved in cognitive functions and may contribute to the negative symptoms and thought disorders in schizophrenic patients.[10-11] As discussed above, the superior efficacy of atypical APDs at treating negative symptoms distinguishes them from the typical agents. Hence, the effects of both classes of APDs on IEG expression in the prefrontal cortical neurons has been examined by several groups. The medial prefrontal cortex comprised of the infralimbic and prelimbic regions shows a third pattern of c-*fos* induction in the forebrain, i.e., clozapine but not haloperidol enhanced c-*fos* expression in this region.[19,31-33,69,70] Even among the newer APDs, those with an atypical profile (e.g., fluperlapine, sertindole) induce c-*fos* in this region.[31,33] However, the effects of remoxipride and risperidone, both touted as atypical agents, on Fos induction in the medial prefrontal cortex remain somewhat controversial.[70] Remoxipride-induced c-*fos* expression was observed by Robertson et al.[31] and Merchant et al.,[70] but not by Deutch and Duman.[69] Similarly, risperidone was not shown to have an effect[31] or produced a robust induction[33] in Fos-like immunoreactivity in the medial prefrontal cortex. The discrepant results may be explained partly by differences in doses and/or routes of administration or differences in the specific regions of the medial prefrontal cortex analyzed by the different groups.[70]

The induction of c-*fos* in the medial prefrontal cortex by clozapine has led to the speculation that it may be an index of prefrontal cortical activation and hence efficacy in the treatment of negative symptoms.[31,69,70] Continued clinical testing of the newer atypical agents in the treatment of the negative symptoms will help test this possibility.

Interestingly, clozapine-induced c-*fos* mRNA expression and Fos immunoreactivity predominates in deep cortical layers suggesting that the glutamatergic corticostriatal neurons may be the primary targets of atypical APDs.[69,70] A small percent of Fos neurons co-express parvalbumin,[69] a calcium-binding protein seen in GABAergic interneurons in the prefrontal cortex. Additionally, some of the c-*fos* mRNA and Fos-containing neurons form two to three columns running across the cortical layers in coronal sections of the infralimbic and prelimbic cortex. Expression of c-*fos* in the cells within these columns is differentially regulated by chronic treatment with APDs (see Section 3.2) and hence they may contribute to the long-term (therapeutic) effects of APDs as discussed below.

2.1.4   Acute Alterations in c-fos Gene Expression in Other Regions of the Brain

In addition to the nucleus accumbens, an induction in Fos expression by APDs is seen in a number of limbic structures. In the lateral septum and the dorsomedial striatum Fos is induced by all APDs, regardless of their clinical profile.[31] As seen in the medial prefrontal cortex, the only exception to this effect was risperidone, which failed to induce Fos immunoreactivity in the lateral septum.[31] Acute treatment with haloperidol and clozapine also induced Fos in the central nucleus of the amygdala.[34] However, like the prefrontal cortex, an induction of Fos selectively by clozapine but not haloperidol is seen in the periventricular and supraoptic nuclei of the hypothalamus,[34] the Islands of Calleja[32] and the thalamic paraventricular nucleus.[34A]

In summary, an induction of c-*fos* expression in the nucleus accumbens, dorsomedial striatum, lateral septum and possibly amygdala is produced by acute administration of a vast majority of clinically tested typical as well as atypical APDs (Figure 1). These data suggest that changes in neuronal activity in these regions may contribute to the therapeutic effects of APDs. On the other hand, the medial prefrontal cortex and the DLSt appear to be selectively targeted, respectively, by atypical APDs and neuroleptics. Alterations in the prefrontal cortical neuronal activity may participate in the efficacy against negative symptoms whereas that in the DLSt may contribute to the EPS liability of neuroleptics (Figure 2).

2.2   Induction of *Zif* 268 Expression

Zif 268 is a member of the zinc finger transcription factor family induced in many cell types by extracellular stimuli[71] that is constitutively expressed at a relatively high level in the cortex as well as the striatal complex[72] (see also Chapter 4). Northern blot analysis showed that both haloperidol and clozapine induced *zif* 268 mRNA in the neostriatum after an acute treatment.[18] However, haloperidol produced a significantly greater induction than clozapine. Quantitative data on haloperidol and clozapine induced *zif* 268 expression by immunohistochemical analysis in the dorsal striatum and the nucleus accumbens was reported recently.[32] However, subregional patterns of expression (if any) within the striatum were not described. Daunais and McGinty[73] recently reported an increase in *zif* 268 mRNA levels in the medial striatum after acute administration of the D2-like antagonist, sulpiride. We have recently observed that unlike the predominantly dorsolateral localization of haloperidol-induced c-*fos* induction, *zif* 268 mRNA expression produced by haloperidol and clozapine is distributed quite

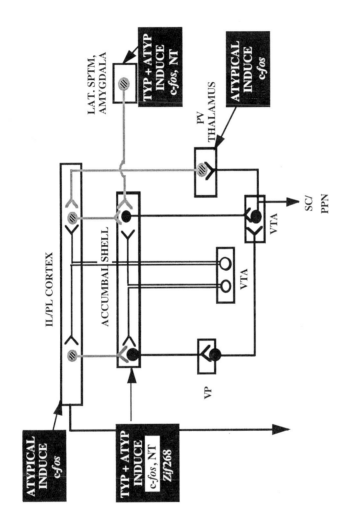

FIGURE 1.

Antipsychotic drug-induced regional gene expression hypothesized to contribute to the therapeutic effects. A simplified schematic diagram of the limbic circuitry shows typical and atypical APD-induced alterations in gene expression. Filled (solid) black neurons are GABAergic projections, neurons with hatched fillings with gray projections represent the excitatory pathways and the open neurons are dopaminergic projections. For the purpose of clarity, the ventral tegmental area (VTA) is separated into two regions to depict the dopamine efferents and GABAergic afferents. The effects that are known to persist after chronic treatment are highlighted. VP = ventral pallidum, PV THALAMUS = paraventricular nucleus of the thalamus, LAT SPTM = lateral septum.

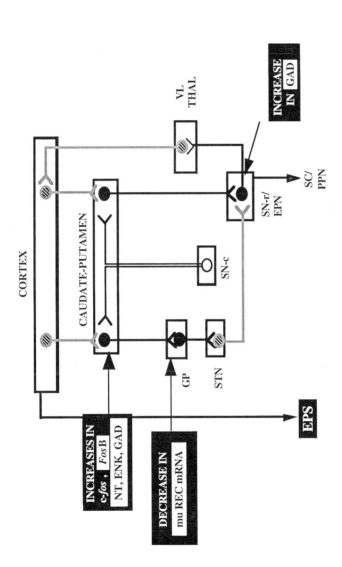

FIGURE 2.

Effects of typical antipsychotic drugs on regional gene expression that may participate in extrapyramidal side effects. This is a simplified schematic diagram of the basal ganglia motor circuitry showing the effects of neuroleptics. As in Figure 1, filled and hatched cell bodies represent, respectively, GABA and excitatory pathways and the open cell bodies are dopaminergic neurons. The effects that are known to persist or be induced after chronic treatment are highlighted. GP = globus pallidus, EPN = entopeduncular nucleus (internal segment of the pallidum in primates), SN-c and SN-r represent the compacta and reticulata of substantia nigra, respectively, VL THALAMUS = ventrolateral thalamus.

uniformly throughout the dorsal striatum (in preparation). The down-stream target(s) of *zif* 268 remain unclear at the present time and its identification is crucial for an understanding of the mechanisms under-lying APD effects. Induction of this transcription factor (if in the same target cells) by both haloperidol and clozapine indicates that it may not be involved in motoric side effects. However, the effects of other typical and new putatively atypical APDs on *zif* 268 expression need to be characterized in order to understand the functional significance of this transcription factor induction.

3. CHRONIC EFFECTS OF APDs

3.1   Alterations in c-*fos* and NT/N mRNA in the Caudatoputamen

Besides the correlation between c-*fos* induction in the DLSt and the clinically observed EPS liability of the APD, the afferent and efferent connectivity of this region demonstrates that it is involved predomi-nantly in regulation of motor functions.[21] Hence, an induction of c-*fos* by neuroleptic drugs with high EPS liability raises the question of whether this effect may be related to the EPS of the neuroleptics. Rodent catalepsy is widely thought to mirror acute EPS in humans.[74] Hence, the obvious question is: Does induction of c-*fos* in the DLSt correlate with catalepsy? Indeed, for the typical APDs the dose response curves for cataleptic effects and Fos induction are correlated. However, atyp-ical APDs even at high doses do not produce the same kind of rigidity that is seen with neuroleptics, partly due to other side effects (e.g., sedation) that interfere with catalepsy measurements. Another con-founding factor to note is that cataleptic scores vary considerably depending on the method of evaluation. This makes it difficult to correlate the dose-response curves for c-*fos* induction and cataleptic responses observed in different laboratories.

Intracerebral injections of neurotensin produce catalepsy in mice.[75] These data suggest that haloperidol-induced neurotensin synthesis (and presumably its release) may contribute to the cataleptic effects of the neuroleptic. Previous studies show that rats develop tolerance to the cataleptic effects of haloperidol upon repeated treatment.[76,77] Hence, if NT or c-*fos* induction by haloperidol were involved in the cataleptic behavior, a tolerance should be observed in NT/N and c-*fos* mRNA induction after chronic haloperidol treatment. Four-week treatment with haloperidol (1 mg/kg/day through subcutaneously implanted osmotic mini pumps) failed to induce c-*fos* in the DLSt and induced NT/N gene expression that was only 50% of that produced by an acute administration (Figure 2).[78] A 24-h wash-out period followed by an acute challenge with haloperidol produced a small but significant c-*fos* mRNA induction but did not reinstate NT/N mRNA response to the level seen after acute treatment.[78] These data indicate that the mechanism underlying c-*fos* tolerance is more rapidly reversible than that leading to NT/N tolerance. A time course of this tolerance indicated that the NT/N gene induction began to develop tolerance to haloperidol administration between 3 and 7 days of commencing the treatment. Interestingly, the rats began to display tolerance also to the cataleptic effects of haloperidol between 3 and 7 days. In fact, an excellent correlation between the cataleptic scores and NT/N mRNA levels was observed in the same animals.[79] These data may have clinical relevance since patients also often develop tolerance to the acute EPS induced by APDs.[80] It is interesting to note that after commencing the chronic haloperidol administration, as the NT/N mRNA levels began to decline, the proenkephalin mRNA levels started to rise (Figure 2). Thus a significant induction in proenkephalin mRNA level was observed first at 7 days after commencing the treatment.[79,81] Accompanying the increases in striatal proenkephalin levels were compensatory declines in the expression of μ-opioid receptor mRNA in the globus pallidus, the terminal region of striatal enkephalin neurons (Figure 2).[81] These data raise two questions: (1) do acute increases in Fos-induced NT/N mRNA by neuroleptic drugs act as a signal for induction of proenkephalin mRNA and its release in the globus pallidus and (2) does enkephalin, in turn, down regulate the expression of NT/N gene? An understanding of peptide-peptide interactions within the striatum could shed light on the physiological relevance of synthesis and storage of multiple neuropeptides and classical neurotransmitters in the brain.

Another evidence of the involvement of the DLSt neurotensin neurons in acute EPS such as parkinsonism is seen in an animal model of Parkinson's disease. Rats with unilateral lesions of the nigrostriatal dopamine projections produced by the neurotoxin, 6-hydroxydopamine, is a commonly used model of biochemical changes occurring in Parkinson's disease.[82] An increase in NT/N mRNA was seen only in the ipsilateral DLSt at 1 and 3 days after 6-hydroxydopamine injections into the neostriatum.[83] However, the lesion-induced increases in NT/N gene expression developed tolerance at a much faster rate than that seen after chronic haloperidol. Thus by day 7 after the lesion, there was no induction of NT/N mRNA levels in the DLSt. Taken together, these data suggest a role of NT neurons in the DLSt in some of the acute motor side effects of neuroleptic drugs.

Antimuscarinic drugs are frequently administered to patients to counteract the EPS of neuroleptic drugs. These agents also block neuroleptic-induced catalepsy. Hence, we examined the effects of atropine (2 mg/kg, i.p.) on acute haloperidol-induced NT/N gene expression. Atropine by itself did not alter NT/N mRNA levels in the DLSt. Neither did it modulate the induction in NT/N mRNA levels produced by haloperidol.[34] These data suggest that the anti-parkinsonian effects of antimuscarinic drugs may be mediated downstream from the changes in NT/N expression. These results are consistent with the failure of antimuscarinic treatment to reduce 6-hydroxydopamine-induced increases in striatal proenkephalin mRNA levels[84] (see also Chapter 7). It should be pointed out that co-administration of scopolamine with haloperidol was shown to attenuate haloperidol-induced Fos-like immunoreactivity in the DLSt.[36] However, a very high dose of scopolamine (5 mg/kg) was tested in this study and hence the relevance of this effect is not clear.

Another long-term consequence of 1 month of continuous haloperidol administration (but not that of clozapine) is a late-emerging, sustained increase in NT/N mRNA in caudal aspects of the ventrolateral striatum.[78] The time course of this effect demonstrated the first detectable effect at 14 days after haloperidol treatment and is maintained even after four months of neuroleptic administration (unpublished observations). These data coupled with involvement of this region in oral movements[85] led us to propose that it may reflect the tardive dyskinesia liability of APDs.

3.2   Induction of c-*fos* mRNA in the Nucleus Accumbens and the Medial Prefrontal Cortex

Whereas continued treatment with neuroleptic agents may produce tolerance in acute EPS of these drugs, the clinical efficacy does not appear to decline with chronic use. Hence, if induction of NT/N mRNA or Fos immunoreactivity were involved in therapeutic effects of these drugs, chronic APD treatment should not lead to tolerance in these effects. In fact, 4 weeks of haloperidol or clozapine treatment induced NT/N mRNA in the accumbal shell similar to that observed after an acute injection.[78] These results are consistent with the report of Kilts et al.[86] who showed persistent elevation in NT immunoreactive content in the ventral striatum after chronic haloperidol treatment. Fos immunoreactivity also remained elevated in the lateral septum after chronic haloperidol and clozapine treatment.[34] In the nucleus accumbens, clozapine-, but not haloperidol-induced Fos levels were maintained after chronic treatment. These data distinguish the APD-induced phenotypic alterations in the accumbal shell and the lateral septum from those in the DLSt neurons and exemplify the benefits of the approach of examining temporal alterations in gene expression to identify anatomical targets involved in the mechanisms of action of APDs.

Chronic (4-week) treatment with clozapine (20 mg/kg through subcutaneously implanted osmotic mini pumps) produced a significantly smaller induction of c-*fos* mRNA in the infralimbic cortex as compared to the acute effect.[70] However, the columns of cells showing c-*fos* mRNA expression after acute administration with clozapine were still distinctly visible after chronic treatment; i.e., the neurons forming these columns did not show the same degree of tolerance as seen in the neurons surrounding the columns. These observations raise the question whether the long-term phenotypic changes maintained by a subgroup of the prefrontal cortical neurons may be involved in the long-term behavioral and biochemical effects of clozapine.[87,88]

3.3   Summary of Chronic Effects of APDs on Gene Expression

Chronic administration of neuroleptics leads to a region-specific tolerance in the expression of NT/N and c-*fos*:

1. The DLSt shows tolerance of c-*fos* and NT/N mRNA induction by haloperidol which coincides temporally with tolerance to cataleptic effects of haloperidol. Chronic neuroleptic administration leads to a sustained increase in proenkephalin and GAD mRNA levels in the neostriatum and a decrease in μ-opioid peptide mRNA in the globus pallidus (Figure 2).

2. In contrast, the nucleus accumbal induction of NT/N mRNA and Fos-like immunoreactivity is maintained after 4 weeks of haloperidol and clozapine administration (haloperidol-induced Fos in the accumbens shows some tolerance).

3. In the medial prefrontal cortex, c-*fos* mRNA shows significant tolerance, however, in neurons organized into columns the degree of tolerance is considerably lower (Figure 1).

3.4   Induction of ΔFosB Expression in the Forebrain

ΔFosB protein lacks a portion of the C terminus sequence present in FosB due to alternative splicing of the IEG, *fosB* (see also Chapter 6). Compared to other IEGs, ΔFosB shows prolonged induction kinetics and hence is an ideal IEG product for studies of chronic APD effects. Interestingly, the regional patterns of induction of ΔFosB by chronic APDs were remarkably similar to those of Fos induction after acute treatment with typical and atypical APDs described above.[89] Thus chronic treatment with haloperidol but not the atypical agents, clozapine and ICI 204,636, induced ΔFosB in the dorsal neostriatum. On the other hand, only the atypical APDs induced ΔFosB in the prefrontal cortex and the lateral septum. Finally, the ventral striatum showed ΔFosB induction by both typical and atypical APDs. Recent studies indicate that Fos regulates the expression of ΔFosB gene. Thus regional alterations in ΔFosB expression by chronic APD treatment may be used as an index of the clinical profile of these drugs.

4. EFFECTS OF DOPAMINE D3- AND D4-SELECTIVE BLOCKERS ON c-*fos* mRNA EXPRESSION

4.1 Classification of Dopamine Receptors

The pharmacological effects of dopamine were thought first to be mediated by two types of receptors: D1, linked to the activation of adenylate cyclase, and D2, not linked or negatively linked to adenylate cyclase.[90] As discussed above, all APDs in the clinical use block the D2-like receptors. Recent molecular cloning studies demonstrate that the D1 and D2 receptors represent subfamilies such that the D1-like receptors are encoded by at least two distinct genes, D1 and D5 (or D1b), and the D2-like receptors encoded by three distinct genes termed, D2, D3 and D4.[91-93] The identification of these novel genes and their distinct anatomical patterns of expression have led to new hypotheses regarding the targets of APDs involved in therapeutic efficacy and side effect profile of APDs.[94,95] The D3 receptor mRNA is seen predominantly in the shell sector of the nucleus accumbens and the Islands of Calleja.[96,97] On the other hand, D4 mRNA is expressed predominantly in the cortex and the hippocampus.[98] Additionally, the endogenous ligand dopamine shows the greatest affinity for the D3 receptor as compared to D2 or D4 receptors expressed in clonal cell lines. On the other hand, the atypical APD, clozapine, shows a small preference for the D4 receptor over D2 and D3 receptors,[93] and at free drug concentrations in the plasma achieved at clinical doses it is thought to preferentially block D4 receptors.[99] These data have led to the suggestion that the D3 and/or D4 receptor may be involved in the therapeutic effects of APDs.[94,95]

4.2 Effects of Dopamine D3 Blockade on c-*fos* and NT/N Gene Expression

To examine the role of dopamine D3 receptors in regulation of the c-*fos* gene in the DLSt, nucleus accumbens and the infralimbic cortex, two pharmacological tools were employed. First, the acute effects of an antagonist, U-99194A, with 20-fold preference for D3 over D2 receptors was examined. The doses were based on the behavioral activation produced by U-99194A.[100] Several groups have now demonstrated that in contrast to the locomotor inhibitory effects of nonselective D2-like antagonists with high affinity for the D2 receptor, the D3-preferring antagonists produce locomotor activation in rats.[100,101] Additionally, D3 knockout mice show hyperlocomotor activity as compared to the wild-type mice.[102] The locomotor activation by D3-preferring antagonists is seen also in animals depleted of dopamine pools leading to the hypothesis that it is the postsynaptic D3 receptor that plays an inhibitory role.[103] At the smallest dose (25 μmol/kg) that produced behavioral activation, U-99194A produced a large increase in c-*fos* mRNA levels in the infralimbic/ventral prelimbic cortex.[70] However, it failed to induce c-*fos* mRNA expression in the nucleus accumbens or the DLSt, nor did it alter NT/N mRNA levels in these subcortical regions.[90] Interestingly, a fourfold increase in the dose produced an induction of c-*fos* mRNA in the infralimbic/ventral prelimbic cortex that was smaller than that produced by the dose of 25 μmol/kg. Additionally, the higher dose also increased c-*fos* and NT/N mRNA in the DLSt in one third of the animals. The DLSt response is likely to be mediated by the D2 receptor, since it is the predominant subtype of D2-like receptors in this region. Even at a single cell level, a recent preliminary report shows the presence of predominantly D2 but not D3 transcripts in striatopallidal neurons,[104] the subtype of medium spiny neurons that express Fos immunoreactivity after haloperidol treatment.[37] Finally, our preliminary double labeling studies indicate co-expression of D2 and c-*fos* mRNA in the DLSt neurons, although co-expression of D3 and c-*fos* mRNA was not examined in this study. Collectively, the dose-dependent modulation of c-*fos* mRNA in the infralimbic/ventral pre-limbic cortex vs. the DLSt by U-99194A has led us to hypothesize that in the medial prefrontal cortex, blockade of D3 receptors may lead to induction of c-*fos* but that of D2 receptors may antagonize the same.[70] An indirect support for this hypothesis is seen also in the effects of remoxipride, an antagonist which binds selectively to only D2 receptors *in vitro* but generates in rats active metabolites with high D3 affinity.[105] Remoxipride also induced c-*fos* mRNA in the infralimbic/ventral pre-limbic cortex at very low doses and lost its ability to produce this effect at higher doses.[70] It should be noted, however, that remoxipride shows high affinity also for the sigma receptors whose contribution to Fos induction remains unclear.

In view of the relatively small amount of D3 transcripts and binding sites in the prefrontal cortex, it is likely that the induction of Fos is mediated by blockade of D3 receptors at a distant site. One likely candidate is the paraventricular thalamic nucleus which sends reciprocal connections to the medial prefrontal cortex[106] and shows Fos induction after clozapine treatment.[34] This region shows exclusive expression of D3 mRNA within the D2 subfamily. On the other hand, a long-loop polysynaptic effect involving the blockade of D3 receptors in the nucleus accumbens may also be involved. Future studies of tracing the site(s) of action of D3 antagonists mediating Fos induction will help understand the physiological role of these receptors.

In the nucleus accumbens, the role of D3 receptors in regulating NT/N mRNA expression remains controversial. Our data with U-99194A showed an induction of NT/N mRNA only at the high dose of 100 μmol/kg but only in those rats who showed induction in the DLSt. The low (and presumably D3-selective) dose neither increased nor decreased NT/N mRNA levels in the accumbal shell.[70] On the other hand, Diaz et al.[107] suggest that D3 blockade may reduce NT/N mRNA in the ventral shell whereas blockade of D2 receptors may increase the same in the dorsal shell. Future availability of compounds with better selectivity for D2 and D3 receptors may help clarify this issue.

With respect to Fos induction in the accumbal shell, a recent preliminary report demonstrated co-expression of D3 mRNA and Fos-IR in the accumbal shell after an acute treatment with clozapine.[67] Additionally, co-administration of 7-hydroxy-DPAT, a D2/D3 agonist, blocked clozapine-induced Fos expression in this region.[108] These data led to the suggestion that blockade of D3 receptors may increase Fos expression in the accumbal shell. However, after haloperidol treatment, neurons that showed the presence of Fos immunoreactivity did not show the same degree of co-expression of D3 mRNA as that seen after clozapine.[67] Hence, the role of D2 and D3 receptors in the induction of Fos in the nucleus accumbens also remains unclear at the present time.

4.3   Effects of Dopamine D4 Blockade on c-*fos* and NT/N Gene Expression

To understand the role of D4 receptors in regulating IEG expression in the forebrain, we characterized the effects of U-101387, a highly selective antagonist with moderately high affinity for only D4 receptors.[109,110] Acute administration of U-101387 potently increased c-*fos* mRNA expression in the infralimbic/ventral prelimbic cortex. Unlike clozapine, which produced c-*fos* mRNA in predominantly pyramidal neurons (66% of all total neurons), U-101387 induced c-*fos* expression in both pyramidal and interneurons.[110] However, like clozapine, the D4 blocker also induced c-*fos* mRNA in neurons organized into columns spanning the cortical layers (see Section 2.1.3). Interestingly, U-101387-induced c-*fos* mRNA expression did not show an absolute dose dependency. Thus the lowest and the highest doses of 0.42 and 42 μmol/kg were efficacious whereas 4.2 μmol/kg produced an effect slightly smaller than that seen with the other doses. The reasons underlying this observation or its significance remains unclear. Recent data demonstrate that the acute combination of U-101387 (21 μmol/kg, i.p.) and amphetamine (2 mg/kg, s.c.) produces additive increases in c-*fos* mRNA levels in the medial prefrontal cortex (submitted).[111] These data indicate that D4 blockade and amphetamine induce c-*fos* mRNA in distinct populations of neurons.

In agreement with the predominantly cortical localization of D4 receptors, U-101387 failed to alter c-*fos* or NT/N expression in the two subcortical regions, the DLSt and the nucleus accumbens, at all doses tested. It is also interesting to note that U-101387 does not produce any overt behavioral alterations seen either with the nonselective D2 antagonists like haloperidol or the D3-preferring antagonist, U-99194A. Hence, the functional implications of c-*fos* induction in the medial prefrontal cortex remain unclear; however, it is likely to alter neuronal activity in the subcortical regions. This was evident in a recent study that showed that co-administration of U-101387 with amphetamine attenuates the induction of c-*fos* mRNA in the dorsomedial striatum produced by the psychostimulant.[112] These data suggest a role of D4 receptors in modulating corticostriatal glutamate output.

4.4   Summary and Conclusions

In the medial prefrontal cortex, blockade of either D3 or D4 receptors appears to induce c-*fos* mRNA expression (Figure 3). It is unknown whether the induction of c-*fos* gene by D3 and D4 blockade is in the same population of neurons. In contrast to the effects of D3 and D4 antagonists, blockade of D2 receptors may prevent the induction of Fos in the infralimbic/ventral prelimbic cortex. This possibility explains the lack of an effect of haloperidol, a potent nonselective D2, D3, D4 blocker, in inducing Fos in this region. Based on the known distribution of D3 and D4 mRNA, the effects of D4 blockade may be direct whereas those of D3 antagonists may be indirectly mediated via D3 blockade at a distant site.

In the caudate-putamen, blockade of D2 receptors may lead to the induction of c-*fos* and NT/N mRNA in the DLSt. Thus the effects of neuroleptics in the DLSt may be mediated by the antagonism of D2 receptors. On the other hand, blockade of D3 and D4 receptors does not seem to produce acute phenotypic changes in this region. It is likely, however, that D4 receptors modulate psychostimulant-induced immediate early genes in the caudatoputamen indirectly by altering the activity of the corticostriatal projections.

In the nucleus accumbens, blockade of D2 as well as D3 receptors may be involved in induction of c-*fos* and NT/N mRNA. It remains to be seen if D2 and D3 receptors target different populations of neurons.

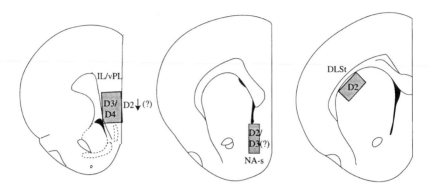

FIGURE 3.
Effects of subtypes of D2-like receptors on regional c-*fos* and neurotensin gene induction. Schematic diagrams of coronal sections of the rat brain depict the hypothesis regarding the involvement of specific subtypes of D2-like receptors in APD effects. The APD-induced c-*fos* expression in the infralimbic/ventral prelimbic cortex (IL/vPL) may be produced by D3 and/or D4 blockade. Antgonism of D2 receptors may prevent this effect. On the other hand, blockade of D2 receptors mediates c-*fos* and NT/N induction in the dorsolateral striatum (DLSt). In the shell sector of nucleus accumbens blockade of both D2 and D3 may increase NT/N gene expression, although it remains controversial as discussed in the text.

As in the caudate-putamen, acute blockade of D4 receptors does not appear to produce significant alterations in c-*fos* or NT/N gene expression, however, the modulatory effects of D4 receptors via cortical efferents remains to be seen.

In conclusion, the APD-induced regional gene expression may be mediated by differential participation of each subtype of the D2 subfamily (Figure 3). The medial prefrontal cortical effects may involve blockade of D3 and/or D4 receptors, the DLSt effects may be mediated predominantly by the D2 receptors and the accumbal responses may involve participation of D2 and/or D3 receptors. Identification of downstream neurochemical targets of Fos in the medial prefrontal cortex and the accumbal shell will help understand the physiological and therapeutic significance of the APD-induced regional patterns of Fos induction.

ACKNOWLEDGMENTS

Some of the studies reviewed in this chapter were performed by the author during her postdoctoral fellowship in the laboratory of Dr. Daniel Dorsa with the much appreciated technical assistance of Anne Kalliomaki. In addition, I am thankful for the collaborative studies with Drs. Margaret Miller and Dorcas Dobie and the help of Ms. RaeLee

Parker. Finally, I am grateful to Mrs. Dawna Feldpausch and Lana Needham for their contribution to the studies of the role of D3 and D4 receptors.

REFERENCES

1. Creese, I., Burt, D.R. and Snyder, S.H., Dopamine receptor binding predicts clinical and pharmacological potencies of antischizophrenic drugs, *Science*, 192, 481, 1976.
2. Seeman, P., Lee, T., Chau-Wong, M. and Wong, K., Antipsychotic drug doses and neuroleptic/dopamine receptors, *Nature*, 261, 717, 1976.
3. Perutka, S.J. and Snyder, S.H., Relationship of neuroleptic drug effects to brain dopamine, serotonin, adrenergic and histamine receptors and clinical potency, *Am. J. Psychiatr.*, 137, 1518, 1980.
4. Meltzer, H.Y., Lee, M.A. and Ranjan, R., Recent advances in the pharmacotherapy of schizophrenia, *Acta Psychiatr. Scand.*, 90 (Suppl. 384), 95, 1994.
5. Kane, J.M. and Freeman, H.L., Towards more effective antipsychotic treatment, *Br. J. Psychiatr.*, 165 (Suppl. 25) 22, 1994.
6. Baldessarini, R.J. and Frankenburg, F.R., Clozapine: a novel antipsychotic drug, *N. Engl. J. Med.*, 324, 746, 1991.
7. Deutch, A.Y., Moghaddam, B., Innis, R.B., Krystal, J.H., Aghajanian, G.K., Bunney, B.S. and Charney, D.S., Mechanisms of actions of atypical antipsychotic drugs: implications for novel therapeutic strategies of schizophrenia, *Schizophrenia Res.*, 4, 121, 1991.
8. Kerwin, R.W., The new atypical antipsychotics. A lack of extrapyramidal side effects and new routes in schizophrenia research, *Br. J. Psychiatr.*, 164, 141, 1994.
9. Dolan, R.J., Functional imaging and the neurobiology of the psychoses, *Neurosciences* 7, 165, 1995.
10. Kotrla, K.J. and Weinberger, D.R., Brain imaging in schizophrenia, *Annu. Rev. Med.*, 46, 113, 1995.
11. Morgan, J.I. and Curran, T., Stimulus-transcription coupling in the nervous system: involvement of the inducible proto-oncogenes fos and jun, *Annu. Rev. Neurosci.*, 14, 421, 1991.
12. Dragunow, M. and Faull, R.L.M., The use of c-*fos* as a metabolic marker in neuronal pathway tracing, *J. Neurosci. Methods*, 29, 261, 1989.
13. Chiu, R., Boyle, W., Meek, J., Smeal, T., Hunter, T. and Karin, M., The c-*fos* protein interacts with c-*jun*/AP-1 to stimulate transcription of the AP-1 responsive genes, *Cell*, 54, 541, 1988.
14. Nakabeppu, Y. and Nathans, D., A naturally occurring truncated form of FosB that inhibits Fos/Jun transcription activity, *Cell*, 64, 751, 1991.
15. Christy, B. and Nathans, D., DNA binding site of growth factor-inducible protein, Zif268, *Proc. Natl. Acad. Sci. U.S.A.*, 86, 8737, 1989.
16. Miller, J.C., Induction of c-*fos* mRNA expression in the rat striatum by neuroleptic drugs, *J. Neurochem.*, 54, 1453, 1990.
17. Dragunow, M., Robertson, G.S., Faull, R.L.M., Robertson, H.A. and Jansen, K., D2 dopamine receptor antagonists induce Fos. and related proteins in rat striatal neurons, *Neuroscience*, 37, 287, 1990.
18. Nguyen, T.V., Kosofsky, B.E., Birnbaum, R., Cohen, B.M. and Hyman, S.E., *Proc. Natl. Acad. Sci. U.S.A.*, 89, 4270, 1992.
19. Robertson, G.S. and Fibiger, H.C., Neuroleptics increase c-*fos* expression in the forebrain, Contrasting effects of haloperidol and clozapine, *Neuroscience*, 46, 315, 1992.

20. Deutch, A.Y., Lee, M.C. and Iadarola, M.J., Regionally specific effects of atypical antipsychotic drugs on striatal Fos expression, the nucleus accumbens shell as a locus of antipsychotic action, *Mol. Cell. Neurosci.*, 3, 332, 1992.

21. Carelli, R.M. and West, M.O., Representation of the body by single neurons in the dorsolateral striatum of the awake unrestrained rat, *J. Comp. Neurol.*, 309, 231, 1991.

22. Merchant, K.M. and Dorsa, D.M., Differential induction of neurotensin and c-*fos* gene expression by typical vs. atypical antipsychotics, *Proc. Natl. Acad. Sci. U.S.A.*, 90, 3447, 1993.

23. Ogren, S.-O., Hall, H., Kohler, C., Magnuson, O., Lindbom, L.-O., Angeby, K. and Florvall, L., Remoxipride, a new potential antipsychotic compound with selective anti-dopaminergic actions in the rat brain, *Eur. J. Pharmacol.*, 102, 459, 1984.

24. Mendelwicz, J., de Bleeker, E., Cosyns, P., Deleu, G., Lostra, F., Masson, A., Martens, C., Parent, M., Peuskens, J., Suy, E., de Wile, J., Wilmotte, J. and Norgard, J., A double-blind comparative study of remoxipride and haloperidol in schizophrenic and schizophreniform disorders, *Acta Psychiatr. Scand.*, 82 (Suppl. 358), 138, 1990.

25. Kohler, C., Hall, H., Magnusson, O., Lewander, T. and Gustafsson, K., Biochemical pharmacology of the atypical neuroleptic remoxipride, *Acta Psychiatr. Scand.*, 82 (Suppl. 358), 27, 1990.

26. Walinder, J. and Holm, A.C., Experience of long-term treatment with remoxipride: efficacy and tolerability, *Acta Psychiatr. Scand.*, 82 (Suppl. 358), 158, 1990.

27. Ahlfors, U.G., Rimon R., Appleberg, B., Hagert, U., Harma, P., Katila, H., Mahlanen, A., Mehtonen, O.-P., Naukkarinen, H., Outakoski, J., Rantanen, H., Sorri, A., Taminnen, T., Tolvanen, E. and Holm, A.-C., Remoxipride and haloperidol in schizophrenia: a double bind multicentre study, *Acta Psychiatr. Scand.*, 82 (Suppl. 358), 99, 1990.

28. Lewander, T., Overcoming the neuroleptic-induced deficit syndrome, clinical observations with remoxipride, *Acta Psychiatr. Scand.* 89 (Suppl. 380) 64, 1994.

29. Krieskott, J., Behavioral Pharmacology of Antipsychotics, in *Handbook of Experimental Pharmacology, Vol. 55*, Hoffmeister, F. and Stille, G., Eds., Springer-Verlag, Berlin, 1980, p. 59.

30. Baldessarini, R.J., Drugs and the Treatment of Psychiatric Disorders, in *Goodman and Gillman's The Pharmacological Basis of Therapeutics*, Goodman, A.G., Rall, T.W., Nies, A.S. and Taylor, P., Eds., Pergamon Press, New York, 1990, p. 383.

31. Robertson, G.S., Matsumura, H. and Fibiger, H.C., Induction of Fos-like immunoreactivity in the forebrain as predictors of atypical antipsychotic activity, *J. Pharmcol. Exp. Ther.*, 271, 1058, 1994.

32. MacGibbon, G.A., Lawlor, P.A., Bravo, R. and Dragunow, M., Clozapine and haloperidol produce a differential pattern of immediate early gene expression in rat caudate-putamen, nucleus accumbens, lateral septum and islands of Calleja, *Mol. Brain Res.*, 23, 21, 1994.

33. Fink-Jensen, A. and Kristensen, P., Effects of typical and atypical neuroleptics on Fos protein expression in rat forebrain, *Neurosci. Lett.*, 182, 115, 1994.

34. Sebens, J.B., Koch, T., Ter Horst, G.J. and Korf, J., Differential Fos-protein induction in rat forebrain regions after acute and long-term haloperidol and clozapine treatment, *Eur. J. Pharmacol.*, 273, 175, 1995.

34a. Deutch, A.Y., Ongur, D. and Duman, R.S., Antipsychotic drugs induce Fos protein in the thalamic paraventricular nucleus; a novel locus of antipsychotic drug action, *Neuroscience*, 66, 337, 1995.

35. Merchant, K.M., Dobie, D.J. and Dorsa, D.M., Expression of the proneurotensin gene in the rat brain and its regulation by antipsychotic drugs, *Ann. N.Y. Acad. Sci.*, 668, 54, 1992.

36. Guo, N., Robertson, G.S. and Fibiger, H.C., Scopolamine attenuates haloperidol-induced c-*fos* expression in the striatum, *Brain Res.*, 588, 164, 1992.

37. Robertson, G.S., Vincent, S.R. and Fibiger, H.C., D1 and D2 dopamine receptors differentially regulate c-*fos* expression in striatonigral and striatopallidal neurons, *Neuroscience*, 49, 285, 1992.
38. Graybiel, A.M., Neurotransmitters and neuromodulators in the basal ganglia, *Trends Neurosci.* 13, 244, 1990.
39. Sugimoto, T. and Mizuno, N., Neurotensin in projection neurons of the striatum and nucleus accumbens, with reference to coexistence with enkephalin and GABA: an immunohistochemical study in the cat, *J. Comp. Neurol.*, 257, 383, 1987.
40. DeLong, M.R., Primate models of movement disorders of basal ganglia circuitry, *Trends Neurosci.*, 13, 280, 1990.
41. Albin, R.L., Young, A.B. and Penney, J.B., The functional anatomy of basal ganglia disorders, *Trends Neurosci.*, 12, 366, 1989.
42. Sonnenberg, J.L., Rauscher, F.J., Morgan, J.I. and Curran, T., Regulation of proenkephalin by Fos and Jun, *Science*, 246, 1622, 1989.
43. Tang, F., Costa, E. and Schwartz, J.P., Increase of proenkephalin mRNA and enkephalin content of rat striatum after daily injection of haloperidol for 2 to 3 weeks, *Proc. Natl. Acad. Sci., U.S.A.*, 80, 3841, 1983.
44. Sabol, S.L., Yoshikawa, K. and Hong, J.-S., Regulation of Methionine-enkephalin messenger RNA in rat striatum by haloperidol and lithium, *Biochem. Biophys. Res. Commun.*, 113, 391, 1983.
45. Angulo, J.A., Cadet, J.L., Woolley, C.S., Suber, F. and McEwen, B.S., Effect of chronic typical and atypical neuroleptic treatment on proenkephalin mRNA levels in the rat striatum and nucleus accumbens of the rat, *J. Neurochem.*, 54, 1889, 1990.
46. Sivam, S.P., Strunk, C., Smith, D.R. and Hong, J.S., Proenkephalin-A gene regulation in the rat striatum: influence of lithium and haloperidol, *Mol. Pharmacol.*, 30, 186, 1991.
47. Mercugliano, M., Saller, C., Salama, A., U'Prichard, D. and Chesselet, M.-F., Clozapine and haloperidol have differential effects on glutamic acid decarboxylase mRNA in pallidal nuclei of the rat, *Neuropsychopharmacology*, 6, 179, 1992.
48. Konradi, C., Kobierski, L.A., Nguyen, T.V., Heckers, S. and Hyman, S.E., The rat cAMP-response-element-binding protein interacts, but Fos protein does not interact, with proenkephalin enhancer in rat striatum, *Proc. Natl. Acad. Sci., U.S.A.*, 90, 7005, 1993.
49. Merchant, K.M., Miller, M.A., Ashleigh, E.A., and Dorsa, D.M., 1991, Haloperidol rapidly increases the number of neurotensin mRNA-expressing neurons in neostriatum of the rat brain, *Brain Res.*, 540: 311-314.
50. Augood, S.J., Kiyama, H., Faull, R.L.M. and Emson, P.C., Differential effects of acute dopaminergic D1 and D2 receptor antagonists on proneurotensin mRNA expression in rat striatum, *Mol. Brain Res.*, 9, 341, 1991.
51. Merchant, K.M., Dobner, P.R. and Dorsa, D.M., Differential effects of haloperidol and clozapine on neurotensin gene transcription in rat neostriatum, *J. Neurosci.*, 12, 652, 1992.
52. Govoni, S., Hong, J.S., Yang, H.Y.-T. and Costa, E., Increase of neurotensin content elicited by neuroleptics in the nucleus accumbens, *J. Pharmacol. Exp. Ther.*, 215, 413, 1980.
53. Merchant, K.M., Bush, L.G., Gibb, J.W. and Hanson, G.R., Dopamine D-2 receptors exert tonic regulation of discrete neurotensin systems of the rat brain, *Brain Res.*, 500, 21, 1989.
54. Frey, P., Fuxe, K., Eneroth, P. and Agnati, L., Effects of acute and long-term treatment with neuroleptics on regional telencephalic neurotensin levels in the male rat, *Neurochem. Int.*, 8, 429, 1986.
55. Levant, B., Bissette, G. and Nemeroff, C.B., Neurotensin, in *Neuropeptides in Psychiatry*, Nemeroff C.B., Ed., APA Press, Washington, 1987, p., 149-168.

56. Kislauskis, E. and Dobner, P.R., Mutually dependent response elements in the cis-regulatory region of the neurotensin/neuromedin N gene integrate environmental stimuli in PC12 cells, *Neuron*, 4, 783, 1990.

57. Merchant, K.M. and Miller, M.A., Coexpression of neurotensin and c-*fos* mRNAs in rat neostriatal neurons following acute haloperidol, *Mol. Brain Res.*, 23, 271, 1994.

58. Robertson, G.S., Tetzlaff, W., Bedard, A., St. Jean, M. and Wigle, N., c-*fos* mediates antipsychotic-induced gene expression in the rodent striatum, *Neuroscience*, 67, 325, 1995.

59. Merchant, K.M., c-*fos* antisense oligonucleotide specifically attenuates haloperidol-induced increases in neurotensin/neuromedin N mRNA expression in rat dorsal striatum. *Mol. Cell Neurosci.*, 5, 336, 1994.

60. Heimer, L., Zahm, D.S., Churchill, L., Kalivas, P.W. and Wohltmann, C., Specificity in the projection patterns of accumbal core and shell, *Neuroscience*, 41, 89, 1991.

61. Zahm, D.S. and Heimer, L., Ventral striatopallidal parts of the basal ganglia in the rat: I. Neurochemical compartmentation as reflected by distribution of neurotensin and substance P immunoreactivity, *J. Comp. Neurol.*, 272, 516, 1988.

62. Nemeroff, C.B. and Cain, S.T., Neurotensin-dopamine interactions in the CNS, *Trends Pharmacol. Sci.*, 6, 201, 1985.

63. Kalivas, P.W., Burgess, S.K., Nemeroff, C.B. and Prange, A.J., Jr., Behavioral and neurochemical effects of neurotensin microinjection into the ventral tegmental area, *Neuroscience*, 8, 495.

64. Erin, G.N., Birkemo, L.S., Nemeroff, C.B. and Prange, A.J., Jr., Neurotensin blocks certain amphetamine-induced behaviours, *Nature*, 291, 73, 1981.

65. Widerlov, E., Lindstrom, L.H., Besev, G., Manberg, P.J., Nemeroff, C.B., Breese, G.R., Kizer, J.S. and Prange, A.J., Jr., Subnormal CSF levels of neurotensin in a subgroup of schizophrenic patients, normalization after neuroleptic treatment, *Am. J. Psychiatr.*, 139, 1122, 1982.

66. Wolf, S.S., Hyde, T.M., Moody, T.W., Saunders, R.C., Weinberger, D.R. and Kleinman, J.E., Autoradiographic characterization of 125I-neurotensin binding sites in human entorhinal cortex, *Brain Res. Bull.*, 35, 353, 1994.

67. Guo, N., Vincent, S.R. and Fibiger, H.C., Dopamine D3 receptors mediate clozapine-induced c-*fos* expression in the forebrain. *Abstr. Soc. Neurosci.*, 21, 1707, 1995.

68. Weinberger, D.R., Berman, K.F. and Zec, R.F., Physiological dysfunction of the dorsolateral prefrontal cortex in schizophrenia. I. Regional cerebral blood flow evidence. *Arch. Gen. Psychiatr.*, 43: 114, 1986.

69. Deutch, A.Y. and Duman, R.S., The effects of antipsychotic drugs on prefrontal cortical Fos expression; cellular localization and pharmacological characterization, *Neuroscience*, 70, 377, 1995.

70. Merchant, K.M., Figur, L.M., and Evans, D.L. Antipsychotic drug-induced expression of c-*fos* mRNA in the medial prefrontal cortex: role of D2 vs. D3 receptors, *Cerebral Cortex*, 6, 1, 1996.

71. Christy, B. and Nathans, D., DNA binding site of the growth factor-inducible protein, Zif268, *Proc. Natl. Acad. Sci. U.S.A.*, 86: 8737, 1989.

72. Herdegen, T., Kovary, K., Buhl, A., Bravo, R., Zimmerman, M. and Gass, P., Basal expression of the inducible transcription factors c-*jun*, JunB, JunD, c-Fos, FosB and Krox-24 in the adult rat brain, *J. Comp. Neurol.*, 354, 39, 1995.

73. Daunais, J.B. and McGinty, J.F., The effects of D1 or D2 dopamine receptor blockade on Zif268 and prodynorphin gene expression in rat forebrain following a short-term cocaine binge, *Mol. Brain Res.*, 35, 237, 1996.

74. Shibata, K., Yamada, K. and Furukawa, T., Possible neurochemical mechanisms involved in neurotensin-induced catalepsy in mice, *Psychopharmacology*, 91, 288, 1987.

75. Janssen, P.A., Niemegeers, C.J. and Schellekens, K.H., Is it possible to predict the clinical effects of neuroleptic drugs (major tranquilizers) from animal data?, *Arzneim Forsch.*, 15, 104, 1965.

76. Lappalainen, J., Hietala, J. and Syvalahti, J., Differential tolerance to cataleptic effects of SCH 23390 and haloperidol after repeated administration, *Psychopharmacology*, 98, 472, 1989.

77. Ossowska, K., Karacz, M., Wardas, J. and Wolfarth, S., Striatal and nucleus accumbens D1/D2 dopamine receptors in neuroleptic catalepsy, *Eur. J. Pharmacol.*, 182, 327, 1990.

78. Merchant, K.M., Dobie, D.J., Totzke, M., Aravagiri, M. and Dorsa, D.M., Effects of chronic haloperidol and clozapine treatment on neurotensin and c-*fos* mRNA in rat neostriatal subregions. *J. Pharmcol. Exp. Ther.*, 271, 460, 1994.

79. Merchant, K.M. and Dobie, D.J., Correlation between tolerance in neurotensin gene expression and cataleptic response in rats following continuous haloperidol treatment, *Abstr. Soc. Neurosci.*, 19, 1364, 1993.

80. Campbell, A. and Baldessarini, R.J. Tolerance to behavioral effects of haloperidol, *Life Sci.*, 29, 1341, 1981.

81. Delfs, J.M., Ellison, G.D., Reisine, T. and Chesselet, M.-F., regulation of µ-opioid mRNA in rat globus pallidus: effects of enkephalin increases induced by short- and long-term haloperidol administration, *J. Neurochem.*, 63, 777, 1994.

82. Ungerstedt, U. and Arbuthnott, G.W., Quantitative recording of rotational behavior in rats after 6-hydroxydopamine lesions of the nigrostriatal system, *Brain Res.*, 24, 485, 1970.

83. Merchant, K.M., Similar to typical antipsychotics, nigrostriatal dopamine lesions induce neurotensin gene expression in the dorsolateral striatum, *Abstracts, Annual Meeting of American College of Neuropsychopharmacology*, 108, 1992.

84. Nisenbaum, L.K., Kitai, S.T. and Gerfen, C.R., Dopaminergic and muscarinic regulation of striatal enkephalin and substance P messenger RNA following striatal dopamine denervation: effects of systemic and central administration of quinpirole and scopolamine, *Neuroscience*, 63, 435, 1994.

85. Delfs, J.M. and Kelly, A.E., The role of D-1 and D-2 dopamine receptors in oral stereotypy induced by dopaminergic stimulation of the ventrolateral striatum, *Neuroscience*, 39, 59, 1990.

86. Kilts, C.D., Anderson, C.M., Bissette, G., Ely, T.D. and Nemeroff, C.B., Differential effects of antipsychotic drugs on neurotensin content of discrete rat brain nuclei, *Biochem. Pharmacol.*, 37, 1547, 1988.

87. Youngren, K.D., Moghaddam, B., Bunney, B.S. and Roth, R.H., Preferential activation of dopamine overflow by chronic clozapine treatment, *Neuroscience Lett.*, 165: 41-44, 1994.

88. Chen, J., Ruan, D., Parades, W. and Gardener, E.L., Effects of acute and chronic clozapine on dopaminergic function in medial prefrontal cortex of awake, freely moving rats, *Brain Res.*, 571, 135, 1992.

89. Vahid-Ansari, F., Nakabeppu, Y. and Robertson, G.S., Contrasting effects of chronic clozapine, seroquel (ICI 204,636) and haloperidol administration on ΔFosB-like immunoreactivity in the rat forebrain, *Neuroscience*, in press.

90. Kebabian, J.W. and Calne, D.B., Multiple receptors for dopamine, *Nature*, 277, 93, 1979.

91. Civelli, O., Bunzow, J.R. and Grandy, D.K., Molecular diversity of the dopamine receptors, *Annu. Rev. Pharmacol. Toxicol.*, 33, 281, 1993.

92. Sokoloff, P., Martres, M.-P., Giros, B., Bouthenet, M.-L. and Schwartz, J.-C., The third dopamine receptor (D3) as a novel target for antipsychotics, *Biochem. Pharmacol.*, 43, 659, 1992.

93. Seeman, P. and VanTol, H.H.M., Dopamine receptor pharmacology. *Trends Pharmacol Sci.*, 15: 264-, 1994.

94. Schwartz, J.-C., Levesque, D., Martres, M.-P. and Sokoloff, P., Dopamine D$_3$ receptor, Basis and clinical aspects, *Clin. Neuropharmacol.*, 16, 295, 1993.

95. Van Tol, H.H.M. and Seeman, P., The dopamine D4 receptor: a novel site for antipsychotic action, *Clin. Neuropharmacol.*, 18, S143, 1995.

96. Landwehrmeyer, B., Mengod, G. and Palacios, J.M., Dopamine D3 receptor mRNA and binding sites in human brain, *Mol. Brain Res.*, 18, 187, 1993.

97. Bouthenet, M.L., Souil, E., Martres, M.P., Sokoloff, P. and Schwartz, J.-C., Localization of D3 mRNA in the rat brain using *in situ* hybridization histochemistry: comparison with dopamine D2 receptor mRNA, *Brain Res.*, 564, 263, 1991.

98. Meador-Woodruff, J.H., Mansour, A., Saul, J. and Watson, S.J., Dopamine receptors and transporters: pharmacology, structure and function, in *Neuroanatomical Distribution of Dopamine Receptor Messenger RNAs*, Niznik, H.B., Ed., Marcel Dekker, New York, 1994, p. 401-414.

99. Seeman, P., Dopamine receptor sequences. Therapeutic levels of neuroleptics occupy D2 receptors, clozapine occupies D4, *Neuropsychopharmacology*, 7, 261, 1992.

100. Waters, N., Lofberg, L., Haadsma-Svensson, S., Svensson, K., Sonesson, C. and Carlsson, A., Differential effects of dopamine D2 and D3 receptor antagonists in regard to release, *in vivo* receptor displacement and behavior, *J. Neural Transm.*, 98, 39, 1994.

101. Griffon, N., Diaz, J., Levesque, D., Sautel, F., Schwartz, J.-C., Sokoloff, P., Simon, P., Constentin, J., Garrido, F., Mann, A. and Wermuth, C., Localization, regulation, and role of the dopamine D3 receptor are distinct from those of the D2 receptor, *Clin. Neuropharmacol.*, 18, S130, 1995.

102. Accili, D., Fishburn, C.S., Drago, J., Steiner, H., Lachowicz, J.E., Park, B.-H., Gauda, E.B., Lee, E.J., Cool, M.H., Sibley, D.R., Gerfen, C.R., Westphal, H. and Fuchs, S., A targeted mutation of the D_3 dopamine receptor gene is associated with hyperactivity in mice, *Proc. Natl. Acad. Sci. U.S.A.*, in press.

103. Waters, N., Svensson, K., Haadsma-Svensson, S.R., Smith, M.W. and Carlsson, A., The dopamine D3-receptor: A postsynaptic receptor inhibitory on rat locomotor activity, *J. Neural Transm.*, 94: 11, 1993.

104. Carter-Russell, Hong, W.-J. and Surmeier, D.J., Coordinated expression of dopamine receptors (D1-D5) in single neostriatal neurons, *Abstr. Soc. Neurosci.*, 21, 1425, 1995.

105. Mohell, N.A., Sallemark, M., Rosqvist, S., Malmberg, A. and Jackson, D.M., Binding characteristics of remoxipride and its metabolites to dopamine D2 and D3 receptors, *Eur. J. Pharmacol.*, 238, 121, 1993.

106. Sesack, S.R., Deutch, A.Y., Roth, R.H. and Bunney, B.S., Topographical organization of the efferent projections of the medial prefrontal cortex in the rat: an anterograde tract-tracing study with Phaseolus vulgaris leucoagglutinin, *J. Comp. Neurol.*, 290, 213, 1992.

107. Diaz, J., Levesque, D., Griffon, N., Lammers, C.H., Martres, M.-P., Sokoloff, P. and Schwartz, J.-C., Opposing roles for D2 and D3 receptors on neurotensin mRNA expression in nucleus accumbens, *Eur. J. Neurosci.*, 6, 1384, 1994.

108. Guo, N., Klitenick, M.A., Tham, C.S. and Fibiger, H.C., Receptor mechanism mediating clozapine-induced c-*fos* expression in the forebrain, *Neuroscience*, 65, 3, 747.

109. Tenbrink, R.E., Bergh, C., Duncan, J.N., Harris, D.W., Lahti, R.A., Lawson, C.F., Rees, S.A., Schlachter, S.K. and Smith, M.W., S(–)-4-[4-[2-(isochroman-1-yl)ethyl]piperazin-1-yl]benzenesulfonamide, U-101387, a selective dopamine D4 antagonist, *J. Med. Chem.*, in press.

110. Merchant, K.M., Gill, G.S., Harris, D.W., Huff, R.M., Lookingland, K.L., Lutzke, B.S., McCall, R.B., Piercey, M.F., Schreur, P.J.K.D., Sethy, V.H., Smith, M.W., Svensson, K.A., Tang, A.H., VonVoigtlander, P.F. and TenBrink, R.E., Pharmacological characterization of U-101387, a dopamine D4 receptor selective antagonist, *J. Pharmacol. Exp. Ther.*, submitted.

111. Garimella, B., Jazayeri, A., Spangler, C.S., Essani, K., Needham, L. and Merchant, K.M., Acute effects of concomitant D4 blockade on amphetamine-induced c-*fos* and NGFI-A mRNA expression in the rat forebrain, submitted.

Chapter **3**

DOPAMINE RECEPTOR FUNCTION: AN ANALYSIS UTILIZING ANTISENSE KNOCKOUT *IN VIVO*

Bao-Cun Sun, Ming Zhang, Abdel-Mouttalib Ouagazzal, Lynn P. Martin, James M. Tepper, and Ian Creese

CONTENTS

0-8493-8550-4/96/$0.00+$.50
© 1996 by CRC Press, Inc.

1. INTRODUCTION

Ever since it was discovered as an independent neurotransmitter in the mid-1950s,[1,2] our understanding of dopamine's (DA) functions has undergone continued development. Dopaminergic systems play an important role in the pathophysiology and therapeutics of many neurological and mental disorders, such as Parkinson's disease, schizophrenia, tardive dyskinesia and drug abuse.[3-7] DA neurons are mainly located in the midbrain from which several major dopaminergic pathways originate. The nigrostriatal pathway originates from the substantia nigra pars compacta (SNC; A9) and terminates principally in the dorsal striatum. Degeneration of SNC DA neurons is the major pathology of Parkinson's disease,[3,4] and DA receptor agonists can ameliorate most of the major symptoms of Parkinson's disease by "replacing" the lost DA. DA neurons in the ventral tegmental area (VTA; A10) project to limbic structures (e.g., the nucleus accumbens, olfactory tubercle, amygdaloid complex, medial prefrontal, cingulate and entorhinal cortices), and comprise the mesolimbic and mesocortical dopamine pathways, respectively. They are most probably involved in the emotional and motivational aspects of behavior and certain aspects of learning and memory. It is believed that disturbances in the mesolimbic and mesocortical DA systems may well contribute to the etiology of schizophrenia. DA receptor antagonists are the standard treatment for psychosis.[5-7]

Because of the clinical importance of the CNS DA pathways, many pharmacological agents have been developed to act on central DA receptors. The existence of two DA receptor subtypes, D1 and D2,* has been proposed — defined by the differential affinities of various dopaminergic drugs and their ability to stimulate (D1) or inhibit (D2) the enzyme adenylate cyclase.[8-10] D1 receptors are located mainly on postsynaptic neurons, while D2 receptors are present on both the soma and dendrites as well as terminals of DA neurons (autoreceptors) in addition to postsynaptic neurons. Although D1 and D2 receptors have opposite effects on adenylate cyclase activity, they often produce synergistic effects in many behavioral and electrophysiological responses. This may be due in part to their association with multiple second messenger systems.

Recent molecular cloning studies, however, have demonstrated that the classification of DA receptors into two subtypes is oversimplified. At least five distinct DA receptor subtype genes have been cloned (D_1, D_2, D_3, D_4, D_5). They have different regional distributions and each is linked to intracellular effectors and/or second messengers.[11,12] Based on the homology of the receptors' amino acid sequences and hence their pharmacological properties, the five subtypes comprise two families: the D1 family consisting of the D_1 and D_5 receptors and the D2 family consisting of D_2, D_3 and D_4 receptors.[13-17] Almost all drugs currently available are only selective between the two families but not selective among members of each family. Furthermore, co-expression of many DA receptor subtypes within the same brain region or even within a neuron makes it difficult to define the functional role of any individual receptor subtype. The conventional solution to this problem requires that selective drugs for each receptor subtype be developed — a difficult and time-consuming task. Our laboratory has recently pioneered a different approach, the use of an *in vivo* antisense "knockout" strategy to analyze the function of each individual DA receptor subtype.[18-22]

2. ANTISENSE STRATEGY

2.1 Mechanism and Design of Antisense Oligodeoxynucleotides for Receptor Knockout

The efficiency and selectivity of the antisense knockout strategy both *in vitro* and *in vivo* have previously been examined extensively,[23-25] but only recently has it been applied to the CNS.[26-28] Typically, an antisense oligodeoxynucleotide will comprise between 15 to 30 bases

---

* Nomenclature: D1 and D2 refer to pharmacologically defined receptor families. D_1, D_2, etc. refer to molecularly defined receptor subtypes.

to ensure absolute selectivity of hybridization to its target mRNA by Watson-Crick base pairing. An oligonucleotide of this length, on a statistical basis, should have only one unique target sequence in the genome and should hybridize well with the target mRNA at body temperature. Longer sequences (e.g., >50 bases) might bind to non-target sequences where partial complementarity occurs and thus decreases the specificity of the knockout. The design of oligodeoxynucleotides as potential inhibitors of gene expression must also avoid the selection of sequences that have a substantial amount of either external or internal complementarity, which can lead to either self-association or intermolecular hairpin formation and will obviously decrease the efficiency of binding to the target mRNA, as well as influence the transport and uptake of the oligomer by cells.

The short synthetic oligodeoxynucleotides are thought to enter cells by receptor-mediated endocytosis or non-selective pinocytosis and then bind to mRNA.[29,30] However, despite the success of using oligodeoxynucleotides to inhibit gene expression *in vitro*, the precise molecular mechanisms involved are still not fully understood — especially as they may relate to neurons. The inhibition of translation of a target mRNA by bound antisense oligodeoxynucleotides might involve the ubiquitous enzyme RNase H, which hydrolyzes the RNA part of RNA-DNA duplex formed through the hybridization process. Alternatively, the formation of the RNA-DNA duplex may serve to block the binding and translocation of ribosomes along the mRNA and thereby prevent the continued synthesis of the target protein. It has been reported that antisense oligodeoxynucleotides which are complementary to the area around the initiation codon of a mRNA have more knockout efficiency than those complementary to other sites upstream or downstream along the mRNA.

Initial attempts to inhibit gene expression utilized natural oligodeoxynucleotides with phosphodiester linkages. However, normal DNA is very sensitive to degradation by cellular nucleases. In order to meet the challenge of using this strategy *in vivo*, much effort has been devoted to modifying the oligodeoxynucleotides chemically to make them nuclease-resistant.[24,31] It must be noted that the levels of nuclease activity in CSF are relatively low and success has been achieved with unmodified oligos. Most modifications have been introduced into the phosphodiester backbone. In our laboratory, we have chosen to use a backbone-modified phosphorothioate "S"-oligo which has increased nuclease resistance and has been shown to still be taken up by cells.[32,33] Continuous intraventricular infusion of the S-oligo at 1.5 nmol/h by implanted osmotic minipump can maintain micromolar concentrations of intact phosphorothioate oligo in the CSF for at least 1 week without obvious toxicity.[34] Thus, the extensive brain penetration and rapid cellular uptake make phosphorothioate oligodeoxynucleotides a good choice for brain receptor knockout. The success of the technique is dependent on the (receptor) protein of interest having a rapid turnover rate. For the dopamine receptors, inferential studies using irreversible antagonists suggest a receptor half life ranging from hours to a few days,[35-37] which is suitable for utilizing the antisense knockout technique.

Our laboratory has applied the antisense knockout strategy in the study of dopamine receptors in the CNS, *in vivo*, in order to determine the behavioral and physiological function of various dopamine receptor subtypes. We have been able to selectively block the synthesis of D_2, D_3 or D_4 dopamine receptors, and thus determine their functions, especially their roles in motor behavior and the regulation of dopamine neuron function.[18-22]

2.2   Advantages and Disadvantages of the Antisense Techniques

Antisense oligonucleotides offer several potential advantages over classical antagonists as research tools. For example, antisense oligonucleotides are easy to design, based on the nucleotide sequence of the gene for the target protein. This avoids the need to synthesize and screen, on many different assays, innumerable compounds to find one that binds selectively and with adequate affinity to the target receptor protein. With the discovery of receptor families for most neurotransmitters, the antisense knockout strategy becomes an even more attractive and perhaps obligatory technique. Using the antisense approach, the critical receptor subtype for a given behavioral or physiological response can be quickly identified. Classical approaches along with more sophisticated molecular modeling approaches based on receptor tertiary structure can then be intensively applied for just that single receptor subtype to design, rationally, a classical pharmacological antagonist that may have therapeutic potential. Although antisense "drugs" appear unlikely in the near future for treating CNS disorders because they do not cross the brain/blood barrier, current synthetic approaches to modify the oligo backbone to make it more nuclease-resistant suggest that peripherally administered oligos may ultimately be developed with central activity if a proper delivery method is found. Novel chemical structures which can also bind selectively to mRNA may also play an important future role.

The antisense strategy contrasts with the classical technique of homologous recombination to produce transgenic knockouts in several aspects. The antisense approach can be applied at any stage of development. Transgenic animals, by definition, lack the protein (receptor) of interest from conception, which may be lethal or induce compensatory mechanisms during development. Using a combination of different antisense oligos, a range of phenotypes can be created. Furthermore, the change in gene expression produced by antisense treatment is a reversible process, allowing for an animal to serve as its own control. The antisense strategy may also ultimately have some therapeutic uses if satisfactory delivery mechanisms can be developed — a very unlikely prospect for transgenic approaches.

Some potential problems have been associated with antisense oligodeoxynucleotides. Among them, the greatest drawback may be that oligodeoxynucleotide treatment often results in an incomplete knockout. To interpret receptor binding measurements of receptor loss after antisense treatment with respect to functional indices of receptor loss, one must consider whether there is a linear relation between agonist binding at a receptor and the physiological effects of this binding. In the case where "spare" receptors exist for a particular response — where only a small fraction of receptors must bind agonist in order to saturate the effector mechanism, incomplete knockout may not result in an equivalent functional loss. Thus one may find that oligodeoxynucleotide treatment, although reducing target receptors, may not affect the functional responses of the system because of a large receptor reserve. Obviously, the lack of a physiological or behavioral effect following antisense knockout sometimes will mask the real relation between an individual receptor and these functions. A number of studies have used the irreversible receptor antagonist (a protein modifying reagent) EEDQ to irreversibly block DA receptors to varying degrees *in vivo* to determine the receptor occupancy / response curves.[38-40] These studies suggest that about 70-90% receptor occupation will produce a full behavioral response. Our DA receptor knockout results are in agreement with these findings.

3. EFFICIENCY AND SPECIFICITY OF ANTISENSE KNOCKOUT OF DOPAMINE D2 FAMILY RECEPTOR SUBTYPES

3.1 The Efficiency of Antisense Knockout

Successful antisense knockouts of the dopamine D2 receptor family (D_2, D_3 or D_4 receptor) in rodent brain have been achieved in our laboratory.[18-22] The D_2 antisense phosphorothioate oligodeoxynucleotide (S-oligo) sequence was 5'-AGGACAGGTTCAGTGGATC-3', which corresponds to D_2 receptor codons 2-8.[15] The random oligo has same base composition but in a random order (5'-AGAACGGCACT-TAGTGGGT-3'). The S-oligos were administered by several routes to the rat brain for various periods of time. Continuous intracerebral intraventricular (i.c.v.) infusion was achieved by using a subcutaneously implanted osmotic minipump (infusion rates 0.5-1 $\mu l/h$). Infusion of S-oligo with lower flow rates directly into certain brain nuclei, such as the substantia nigra or striatum, was applied by an external syringe pump (0.1–0.2 $\mu l/h$). Multiple injections into the ventricle or brain nuclei were also tried. Each delivery method decreased the D2 receptor levels although with different knockout efficiencies and regional effectiveness. The i.c.v. administration of D_2 antisense S-oligo (10 $\mu g/\mu l$ × 1 $\mu l/h$) for 72 h resulted in significant loss in D2 receptor density without a change in the affinity of the remaining receptors measured in homogenate binding assays (Figure 1). Autoradiographic studies indicated about a 50% decrease in the striatum, 70% decrease in nucleus accumbens and 49% decrease in substantia nigra binding of ^{3}H-spiperone, a radioligand which binds to all D2 family receptor subtypes, suggesting widespread diffusion of the oligo throughout the brain. Local unilateral intranigral infusion of D_2 antisense oligo (10 $\mu g/\mu l$ × 0.1 $\mu l/h$ × 6 days) also effectively decreased the specific ^{3}H-spiperone binding by about 50% in the infused SNC, while the binding in contralateral SNC or ipsilateral striatum was not significantly changed by the infusion (Figure 2).

We have also achieved success in using antisense S-oligo to reduce D_3 and D_4 receptors in rat brain. For D_3 receptors in the nucleus accumbens and islands of Calleja, autoradiography with the D_3 agonist ligand ^{3}H-7-OH-DPAT indicated about a 40% decrease in both regions after 3 days' administration (10 $\mu g/\mu l$ × 1 $\mu l/h$, i.c.v.). Intranigral D_3 antisense infusion selectively decreased ^{3}H-7-OH-DPAT binding but not ^{3}H-spiperone binding in the infused SN. One problem appeared when we tried to detect D_4 receptors in the brain, since thus far there is no selective radioligand to label and characterize the D_4 receptors. Previous results have showed that YM-09151-2 and spiperone have high affinity for D_2, D_3 and D_4 receptors, while raclopride has high affinity to D_2 and D_3 receptors but much lower affinity for D_4 receptors.[41] Thus we labeled D_4 receptors by either ^{3}H-YM-09151-2 or ^{3}H-spiperone in the presence of 300 nM raclopride (close to saturating concentration for D_2 and D_3 receptors only). The preliminary results also showed a D_4 receptor decrease after D_4 antisense treatment.

3.2 The Specificity of Antisense Knockout

The major strength of the antisense strategy lies in the specificity it can provide in arresting protein synthesis. The importance of assuring the specificity of the antisense knockout can never be overemphasized in studying a complex system such as the brain. Four oligodeoxynucleotide controls can be used:

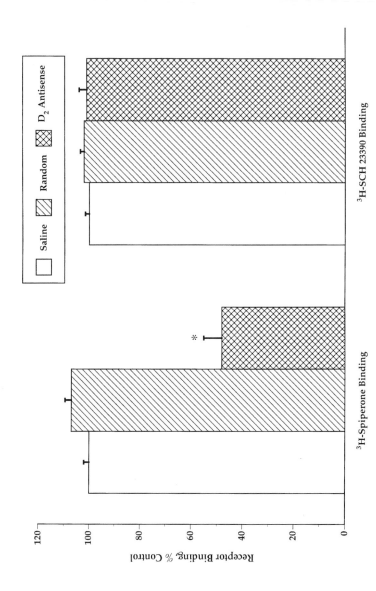

FIGURE 1.

Effects of D_2 antisense S-oligo infusion (10 µg/µl × 1 µl/h × 3 days, i.c.v.) on striatal DA receptors. D_2 antisense treatment selectively decreased D2 receptor (^{3}H-spiperone) binding, but not D1 receptor (^{3}H-SCH 23390) binding. Random S-oligo had no effect on either striatal D1 or D2 receptors. (*p <0.01 compared to saline and random.)

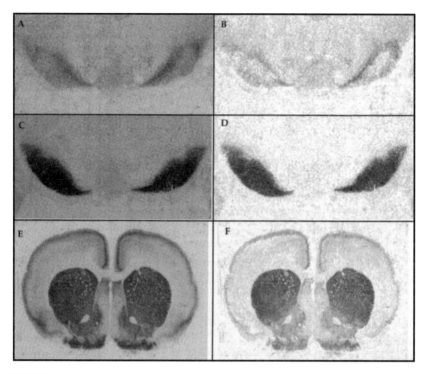

FIGURE 2.
Autoradiogram showing that unilateral infusion of D_2 antisense S-oligo (10 µg/µl × 0.1 µl/h × 6 days) into the left substantia nigra (SN) significantly decreased D2 receptors (^{3}H-spiperone total [A] and specific [B] binding) in the infused SN, while D1 receptors (^{3}H-SCH 23390 total [C] and specific [D] binding) did not change. D_2 receptors in the striatum did not change after the unilateral intranigral D_2 antisense oligo infusion ([E] D2 total binding; [F] D2 specific binding.)

1. The same bases in the "sense" configuration. This has the problem of potentially hybridizing to another specific but unrelated sequence to which it is "antisense."

2. "Random" oligodeoxynucleotides containing the same bases as in the antisense sequence but in a scrambled order.

3. Mismatch oligos in which 3 or 4 bases are mismatched, a number sufficient to reduce the "melting" temperature of the duplex, thus preventing its formation *in vivo*. Because none of these control oligodeoxynucleotides has sequence complementary to the target mRNA, they should not affect the target protein synthesis.

4. A different antisense sequence directed elsewhere along the mRNA of interest which should thus have the same specific antisense knockout of the target protein but different potential non-specific targets.

All control sequences must be checked in Genebank to ensure that they are not antisense for another known sequence. However, only a small fraction of the entire genome has been sequenced to date and thus "specific," non-specific effects may occur. It is preferable at this early stage in the development of the antisense strategy to use all four possible controls until the technique is better understood. However, this is very expensive and time consuming. At this point in time, in our studies on DA receptor antisense knockouts, we have used random oligodeoxynucleotides as the control and found that administration of the random oligodeoxynucleotides failed to reduce DA receptor binding or produce the behavioral effects of the antisense sequence. The selective actions, both behavioral and on radioligand binding, of each specific receptor antisense oligo that we have used also argue for selectivity of action (see below). None of the oligos induced "non-specific" toxicity although S-oligos have been reported to produce non-specific toxicity.[26] This may have been the result of elemental sulfur non-specifically bound to the oligo as a byproduct of a synthesis protocol. Toxic effects may also occur if a number of short oligo sequences are also present because of a poor synthesis protocol or lack of oligo purification. Some sequences of bases within an oligo, such as GGGG, may also be toxic and should be avoided.[42]

The specificity of D_2 antisense knockout was supported by several observations. To determine whether neuronal death might account for the striatal D_2 receptor decrease, we measured striatal muscarinic receptors, a proportion of which are co-expressed on striatal neurons with D_2 receptors,[43] 5-HT_2, D_1 and D_3 receptors after i.c.v. infusion of D_2 antisense S-oligo. None of these receptors was reduced by the D_2 antisense treatment (Figure 1). These are all G-protein-linked receptors and should show similar intracellular processing mechanisms.

In the study of i.c.v. infusion of D_3 antisense S-oligo, we found that ^{3}H-spiperone binding, which labels D_2, D_3 and D_4 receptors, was not reduced in the dorsolateral striatum, an area devoid of D_3 and D_4 mRNA. However, spiperone binding in nucleus accumbens was reduced by about 20%. Considering the fact that D_3 receptor mRNA is expressed predominantly in the nucleus accumbens,[16] it is reasonable to assume that the 20% decrease in ^{3}H-spiperone binding may be due to the loss of D_3 receptors, in keeping with the 40% loss of ^{3}H-7-OHDPAT binding in the nucleus accumbens, while D_2 receptors in the striatum were unaffected.

After unilateral supranigral D_2 or D_3 antisense infusion, there was a selective decrease in specific ^{3}H-spiperone or ^{3}H-7-OHDPAT binding, respectively, but D1 receptors labeled by ^{3}H-SCH 23390 were not affected (Figure 2). In addition, D_2 receptors in the striatum were not affected by intranigral D_2 antisense infusion indicating that there is no retrograde uptake of antisense by the nerve terminals of striatonigral

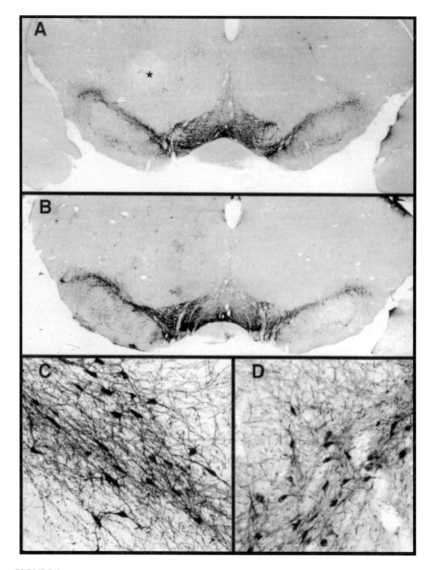

FIGURE 3.
Tyrosine hydroxylase (TH) staining after unilateral infusion of D_2 antisense S-oligo
($10 \mu g / \mu l \times 0.1 \mu l / h \times 6$ days) into the left substantia nigra (SN). (A) D_2 antisense infusion
appeared to increase the TH activity in the infused SN. Note that the infusion site (*)
was just above the left SN. (B) an adjacent slice. (C) Higher power of the infused SNC.
(D) Higher power of the untreated SNC.[22]

neurons (Figure 2).[21,22] Tyrosine hydroxylase immunostaining showed
that the SNC DA cells on the infused side were as healthy as on the
contralateral side (Figure 3),[22] arguing strongly against any non-specific
toxic effects.

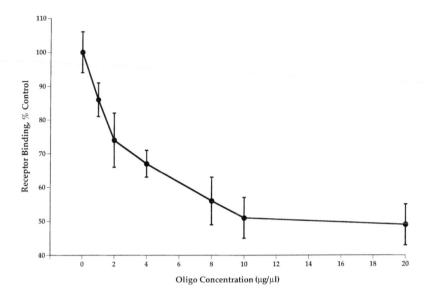

FIGURE 4.

D_2 antisense S-oligo dose-dependent knockout of ^{3}H-spiperone binding in the striatum. D_2 antisense was administered i.c.v. (1 µl/h for 3 days) at various concentrations.

3.3   Effects of Dose and Time on Antisense Knockout

Antisense approaches involve a balance between two opposing factors: the extent of knockout and the specificity of knockout. Because the hybridization between the antisense oligodeoxynucleotide and the target mRNA is probably governed by the extent of secondary structure formation in the target mRNA, it is relatively inefficient. Thus a high degree of hybridization requires an excess of antisense probe over its sense counterpart in the target mRNA. However, at high levels of antisense probe, cross-hybridization to related sequences might occur, limiting the specificity of the approach. Also non-specific neurotoxicity has been documented in several reports when excessive amounts of oligodeoxynucleotide has been administered. To determine the optimal range of dosage of antisense S-oligo, we administered various concentrations (1-20 µg/µl × 1 µl/h) of D_2 antisense S-oligo for 3 days to decrease D_2 receptors. We found that the antisense knockout is somewhat dose-dependent (Figure 4). At a concentration of 1 µg/µl, D_2 receptors were reduced by 14%. At a concentration of 20 µg/µl, the reduction was 58%, slightly higher than at 10 µg/µl. At 20 µg/µl, we did not observe any evidence of significant behavioral toxicity.

One key aspect of protein depletion by antisense approaches is the rate of target protein turnover. Antisense oligodeoxynucleotide only hybridize with the target mRNA so that the synthesis of new receptor

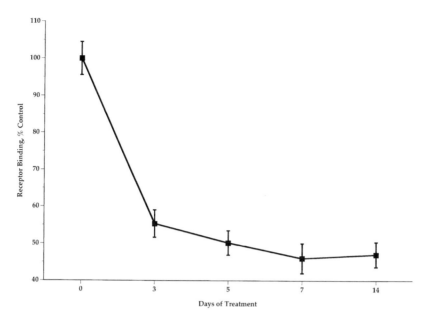

FIGURE 5.

Time couse of the effect of D$_2$ antisense S-oligo treatment (10 µg/µl × 1 µl/h, i.c.v.) on ^{3}H-spiperone binding in the striatum.

protein is blocked. However, the existing receptor protein is not destroyed by the oligodeoxynucleotide. Even with 100% block of protein synthesis by the formation of antisense oligodeoxynucleotide and target sense mRNA duplex, depletion of the target protein will only occur as the remaining pool of the previously synthesized protein is degraded. Therefore, protein degradation rate is an important consideration for transient antisense knockout since several days may be required for depletion of the target protein to produce optimal knockout for functional studies. Degradation is most probably under the control of a number of different processes and therefore may vary across different experimental conditions and even brain regions.[35-37]

Previous studies using the irreversible alkylation of D1 and D2 receptors suggest that the half time for recovery of D2 receptors is 8–160 h, depending on several factors.[35-37] To determine the effect of length of time of antisense treatment on the knockout, we administered 10 µg/µl D$_2$ antisense oligodeoxynucleotide at a delivery rate of 1 µl/h for 1, 3, 5 and 7 days, respectively. Not surprisingly, we found that the knockout is a time-dependent process. Still, even when the treatment was prolonged to 7 days, the "apparent" reduction in D$_2$ receptor density was still not 100% (Figure 5). Because ^{3}H-spiperone also labels D$_3$ and D$_4$ receptors, it is likely that at least some of this residual binding is to these other dopamine receptor subtypes in the striatum.

4. EFFECT OF ANTISENSE KNOCKOUTS OF INDIVIDUAL SUBTYPES OF D2 RECEPTOR FAMILY

Many studies suggest a direct role for CNS DA systems in motor behaviors.[44-47] Clinical observations suggested that some aspects of the movement disorder suffered by Parkinson's disease patients could be alleviated by the administration of DA agonists, while chronic administration of some DA antagonists, as a treatment for schizophrenia, resulted in side effects that mimicked the motor deficits of Parkinsonism. Laboratory studies have found that acute administration of DA antagonists induces hypoactivity and catalepsy, while the nonselective dopamine D1/D2 receptor agonist, apomorphine, and the psychostimulant agent, amphetamine, which releases dopamine, induce hyperactivity and stereotyped behaviors. These observations suggest that increases in dopaminergic tone facilitate motor behavior and vice versa. The search for these neurological substrates, more specifically, the DA receptor subtypes involved in these motor behavioral effects and their CNS locations has inspired many studies in which more selective approaches have been attempted. In the following section we will discuss the effects of antisense knockout of DA receptor subtypes on some aspects of dopamine-related behaviors and neurophysiology of DA neurons.

4.1   Effects of Antisense Knockout of D_2 Receptors

Northern blot analysis and *in situ* hybridization histochemistry have established that the gene for D_2 receptors is abundantly expressed in the brain.[11,15] The areas of highest expression in the brain correspond to major dopaminergic projection areas such as the caudate-putamen, nucleus accumbens and olfactory tubercle. About 50% of the medium-sized cells (most of them are enkephalinergic neurons) as well as large-diameter cells (likely to be cholinergic interneurons) in the striatum have been observed to express D_2 receptor mRNA. D_2 receptor mRNA is also found in dopaminergic cell bodies within the SNC and VTA, indicating a presynaptic (autoreceptor), as well as a postsynaptic, role for the D_2 receptors.

4.1.1   Effects of Intraventricular D_2 Antisense Infusion

To determine the general role of D_2 receptors in the brain, D_2 antisense or random S-oligo (10 mg/μl × 1 μl/h) was infused i.c.v. for 3 days. During antisense treatment, the rats gradually developed a strong cataleptic response (Figure 6). Spontaneous locomotor activity was also

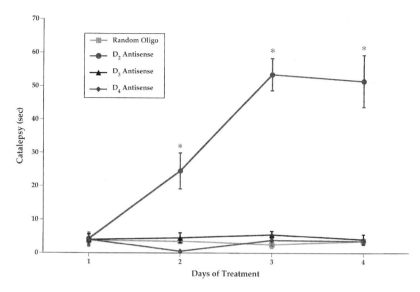

FIGURE 6.

Effects of D_2, D_3 and D_4 antisense treatment ($10~\mu g/\mu l \times 1~\mu l/h$ i.c.v. $\times 3$ days) on catalepsy. D_2 antisense oligo-treated rats were significantly more cataleptic than D_3, D_4 or random oligo-treated rats from day 2 following the onset of infusion (n = 6 in each group, *p <0.01, ANOVA).

reduced in the D_2 antisense treated rats (Figure 7). Furthermore, D_2 antisense treatment decreased the D2 agonist quinpirole-induced loco-motor activity (Figure 8). In contrast, the grooming behavior induced by the D1 receptor agonist, SKF 38393, was not affected. These results suggest that D_2 receptors are involved in the regulation of motor function and may mediate an excitatory effect on motor activity. However, D_2 antisense treatment did not affect the stereotypy and locomotor activity induced by the DA releaser, d-amphetamine (Figures 7 and 9). The different effects of D_2 antisense on the behavioral responses induced by quinpirole and d-amphetamine may be due to the different mechanisms of actions of the two drugs. Quinpirole is a D2 selective agonist and increases locomotor activity by directly activating postsyn-aptic D2 receptors, while the effects of d-amphetamine result from an increase of DA release, which activates D1 as well as D2 receptors to produce behavioral hyperactivity. The D_2 antisense oligo may thus influence DA release (via autoreptors), in addition to simply reducing postsynaptic D_2 receptors.

Weiss et al. have applied multiple injections of a D_2 receptor anti-sense oligodeoxynucleotide into the cerebral ventricles to study dopam-ine D_2 receptors in mice.[48,49] (See Chapter 8.) The model utilized mice with

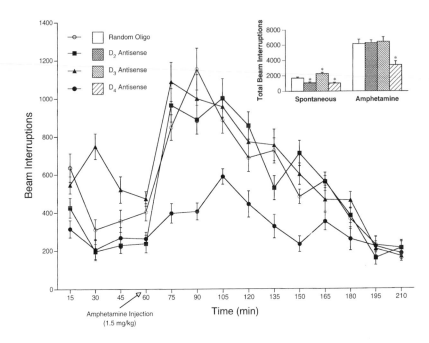

FIGURE 7.
Effects of D_2, D_3 and D_4 antisense treatment (10 μg/μl × 1 μl/h i.c.v. × 3 days) on spontaneous and d-amphetamine-induced locomotor activity. d-Amphetamine (1.5 mg/kg) was injected on day 4, 60 min after rats were placed in the cage. Locomotor activity was measured by photocell beam interruptions. Inset panel: total beam interruptions during habituation period (60 min) and amphetamine stimulation period (150 min). Data represent mean +SEM (n = 6 in each group, $^*p < 0.05$, ANOVA).

unilateral 6-hydroxydopamine lesion of the striatum. Acute administration of dopamine D1 or D2 receptor agonists induced contralateral rotational behavior in these animals. This behavior is dependent on denervation-induced supersensitivity of DA receptors within the lesioned striatum and subsequent imbalance in agonist-induced stimulation of motor activity. D_2 antisense treatment prevented the occurrence of D2-agonist quinpirole-induced rotation in the 6-OHDA-lesioned mice. In contrast, the rotations induced by D1 agonist, SKF 38393, or muscarinic cholinergic agonist, oxotremorine, were not changed by D_2 antisense treatment. In this experiment, the antisense treatment only produced a minor reduction in the D_2 receptor protein in the lesioned striatum, although the decrease was statistically significant. The lower level of receptor losses they obtained may be due to the lower doses of the oligo, species difference in receptor turnover and/or the discontinuous administration mode of the antisense oligodeoxynucleotide. They have hypothesized that the marked behavioral effects indicate a small pool of functional, rapidly turning-over receptors.

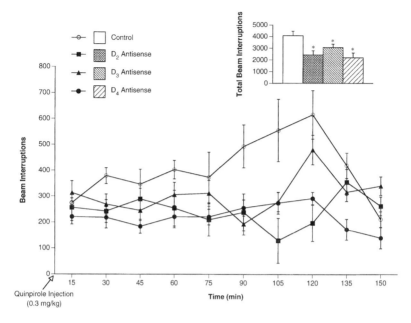

FIGURE 8.

Effect of DA receptor antisense oligodeoxynucleotides on quinpirole-induced locomotor activity. Rats were infused i.c.v. with D_2, D_3, or D_4 antisense oligodeoxynucleotides ($10\ \mu g/\mu l \times 1\ \mu l/h$) for 3 days. Quinpirole ($0.3\ mg/kg$,) was injected on day 4. Locomotor activity was measured by photocell beam interruptions. Inset panel: total beam interruptions induced by quinpirole during the 150-min observation period. Data represent mean +SEM ($n = 6/$group, $*p < 0.05$, ANOVA).

4.1.2 Function of Striatal Postsynaptic D_2 Receptors

To determine the function of D_2 receptors located in specific brain regions and specific subcellular or synaptic (e.g., presynaptic or postsynaptic) locations, D_2 antisense oligo was administered directly into the brain. One of the greatest advantages of antisense knockout over conventional receptor antagonist blockade is that the antisense oligo can selectively knock out one kind of postsynaptic receptor without affecting the same type of receptor located presynaptically in the same region, or vice versa. For example, administration of D_2 antisense directly into the striatum will selectively knock out the postsynaptic D_2 receptors residing on the striatal neurons, which are synthesized by the intrinsic striatal neurons, without affecting the presynaptic D_2 autoreceptors located on the terminals of nigrostriatal DA neurons, which are synthesized by cell bodies of the nigral DA neurons 5–7 mm away. Conversely, intranigral administration of the D_2 antisense oligo will selectively reduce the D_2 autoreceptors located on soma and dendrites and the presynaptic D_2 autoreceptors which are transported to the terminals in the striatum without disrupting the striatal postsynaptic D_2 receptors.

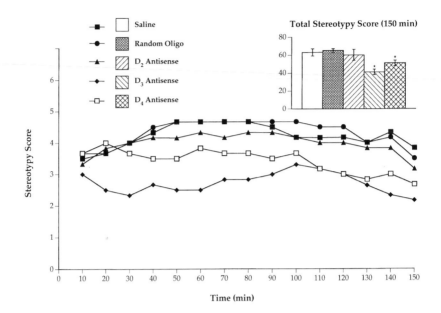

FIGURE 9.

Effects of D_2, D_3 and D_4 antisense treatments ($10 \, \mu g/\mu l \times 1 \, \mu l/h$ i.c.v. $\times$ 3 days) on d-amphetamine-induced stereotypy. Amphetamine ($5 \, mg/kg$) was injected on day 4. Stereotypy was evaluated using a scale created by Creese and Iversen:[40] 1 — no movement; 2 — intermittent locomotor activity with sniffing; 3 — continuous activity and sniffing; 4 — intermittent stereotypy (i.e., pronounced sniffing and rearing); 5 — continuous stereotypy over a wide area (head down and repetitive movements of the head and limbs); 6 — pronounced continuous stereotypy over restricted area. Inset panel shows total stereotypy scores during the 150-min observation following d-amphetamine injection. Data represent mean ±SEM (n = 6 in each group, *p <0.05, Mann-Whitney U-test after significant difference detected by Kruskal-Wallis test).

Unilateral, multiple injections or continuous infusion of D_2 antisense into the striatum was used to study the function of postsynaptic D_2 receptors in striatum. Rats that received unilateral intrastriatal injection of D_2 antisense oligo (1–20 μg) twice a day for 3 days showed a dose-dependent ipsilateral rotational response (toward the injected side) to the DA agonists, apomorphine (2 mg/kg, i.p.) or quinpirole (0.5 mg/kg, i.p.; Figure 10), although they did not show significant spontaneous rotation. These results reinforce the idea that postsynaptic D_2 receptors in the striatum are involved in the modulation of motor behavior.

4.1.3 Function of Somatodendritic and Terminal D_2 Autoreceptors of the Nigrostriatal DA Neurons

The autoreceptors located on nigral DA neuron somatodendritic area and the striatal terminals were originally classified pharmacologically as

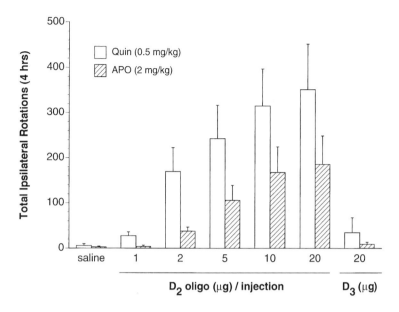

FIGURE 10.

Ipsilateral rotation induced by apomorphine (APO) (2 mg/kg, i.p.) and quinpirole (Quin) (0.5 mg/kg, i.p.) in rats which had received unilateral intrastriatal D_2 or D_3 antisense oligodeoxynucleotide injections twice a day for 3 days at various doses. The D_2 antisense oligodeoxynucleotide treatment induced a dose-dependent increase in ipsilateral rotations in rats when challenged with apomorphine or quinpirole, while the D_3 antisense oligodeoxynucleotide treatment failed to induce significant ipsilateral rotations. Data represent mean ±SEM (n = 5-7).

D_2 receptors. The somatodendritic autoreceptors regulate the firing of DA neurons, and activation of these receptors inhibits the firing of the DA neurons.[50] Activation of DA terminal autoreceptors inhibits the synthesis and release of DA.[51] To determine the contribution of the D_2 receptor as opposed to the D_3 receptor (whose mRNA may also be expressed in DA neurons) in modulating dopaminergic neurotransmission, we infused D_2 antisense S-oligo directly into the SN (10 μg/μl × 0.1 μl/h) for 6 days to specifically block the synthesis of the D_2 DA autoreceptors. Extracellular single unit recording from SNC DA neurons *in vivo* was performed after intranigral infusion.

Extracellular single unit recording showed that D_2 antisense oligo infusion did not significantly affect the spontaneous firing activity of SNC DA cells; the firing rate and firing pattern in anesthetized rats remained in the normal range after treatments (Table 1). However, the inhibition of SNC DA cell firing induced by the DA receptor agonist apomorphine (i.v.) was significantly attenuated by D_2 antisense infusion (Figures 11 and 12).[22] D_2 antisense infusion shifted the dose-response curve of apomorphine to the right (Figure 12). Apomorphine

TABLE 1.

Effects of Unilateral Nigral Infusion of D_2 Antisense
Oligodeoxynucleotide ($10 \ \mu g / \mu l \times 0.1 \ \mu l / h \times 6$ days)
on the SNC DA Neuron Firing Characteristics

Treatment	Control	D_2 Antisense
Basal firing rate	3.58 + 0.20	4.10 + 0.27
(spikes/s)	(n = 52)	(n = 55)
Coefficient of	0.42 + 0.03	0.44 + 0.03
variation	(n = 52)	(n = 52)
Mean antidromic	1.66 + 0.18	1.06 + 0.17[a]
threshold (mA)	(n = 28)	(n = 18)
% full antidromic	19.62 + 3.27	39.61 + 6.37[b]
spikes	(n = 20)	(n = 18)

[a] $p < 0.05$.
[b] $p < 0.01$ as compared to control, ANOVA.
Data from Reference 22.

($8–128 \ \mu g / kg$, i.v.) always totally inhibited the firing of DA neurons in untreated, random oligo treated or in the contralateral SN, while in D_2 antisense treated SN, apomorphine usually could not completely inhibit the firing, even at extraordinarily high doses ($512–2048 \ \mu g / kg$).

The response of nigrostriatal DA neurons to antidromic stimulation of the striatum was performed in antisense-treated rats to evaluate somatodendritic and axon terminal excitability.[51,52] Appropriate striatal stimulation usually induces "small" antidromic spikes consisting of only an initial segment spike and sometimes produces "big" full spikes consisting of both an initial segment and a somatodendritic spike. The proportion of striatal-evoked antidromic responses consisting of full spikes reflects the somatodentric excitability.[52] The current intensity (threshold) necessary to elicit antidromic responses reflects the terminal excitability.[51] Intranigral D_2 antisense infusion increased the proportion of full spike responses and decreased the antidromic threshold, indicating an increase in somatodendritic and terminal excitability, respectively (Figure 11, Table 1). On the contrary, the striatal stimulation-induced inhibition (mediated by GABA) on the DA neurons was not changed after the infusion.

Tyrosine hydroxylase (TH) staining indicated a potential increase in TH activity in the D_2 antisense infused SN without any histological abnormality (Figure 3). This suggests that tyrosine hydroxylase synthesis may be increased after knockout of D_2 autoreceptors.

Behavioral observation demonstrated that the rats receiving unilateral intranigral infusion of D_2 antisense S-oligo showed spontaneous contralateral rotations, away from the infusion side, from the second day of infusion. This suggests an increased DA release in the ipsilateral striatum as a result of D_2 receptor knockout. When challenged with the

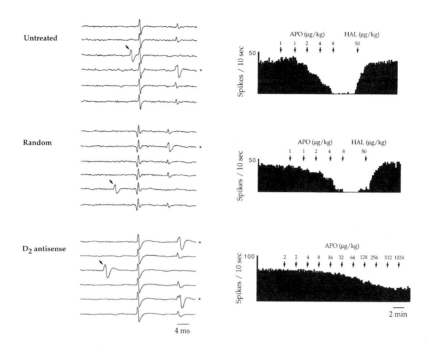

FIGURE 11.

Effects of unilateral intranigral D_2 antisense infusion on the firing activity of nigrostriatal DA neurons. Left panel illustrates the antidromic responses of nigrostriatal DA neurons to striatal stimulation. Full spikes are indicated by asterisks and collisions with spontaneous spikes are indicated by arrows. Right panel illustrates the firing rate histograms. D_2 antisense treatment significantly attenuated the apomorphine (APO)-induced inhibition of DA neurons. 2048 µg/kg (cumulative dose, the number above each arrow indicates dose administered i.v. at that time) of apomorphine still did not completely inhibit the firing rate in a D_2 antisense treated rat, while in an untreated or random oligo treated rat, 16 µg/kg (cumulative) of apomorphine totally inhibited the firing.[22]

DA agonists, apomorphine or quinpirole, no rotational behavior was induced in these rats, indicating that the postsynaptic DA receptors in the striatum were not affected by intranigral antisense infusion. These results suggest that D_2 receptors exhibit a tonic inhibitory role on DA release from nigrostriatal nerve terminals in the awake rat, as has been suggested previously in anesthetized rats.[51]

4.2 Effects of Antisense Knockout of D_3 Receptors

D_3 receptor mRNA is much less abundant and shows a more restricted distribution than D_2 receptors in the brain.[11,16] D_3 receptors are expressed predominantly in limbic brain areas including the olfactory tubercle, nucleus accumbens, islands of Calleja and hypothalamus.

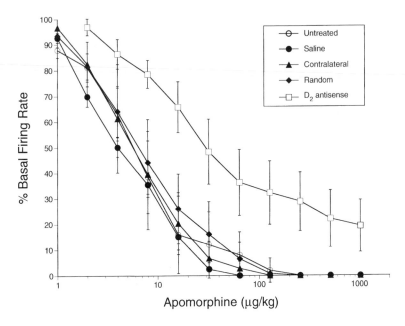

FIGURE 12.
Dose-response curves for apomorphine-induced inhibition of nigrostriatal DA neuron firing after unilateral intranigral D_2 antisense infusion (10 μg/μl × 0.1 μl/h × 6 days). D_2 antisense infusion shifted the curve to the right. Data represent mean ±SEM (n = 6-12).

Low levels of expression are also observed in the caudate-putamen and certain cerebral cortical regions. The almost exclusive localization of the D_3 receptor to the limbic system has led to the hypothesis that D_3 receptor may mediate dopaminergic control of emotional, endocrine and cognitive processes, thus, it may be relevant to antipsychotic action of neuroleptics. D_3 receptor mRNA has also been detected in the SN, indicating that D_3 receptors may also serve as autoreceptors.

4.2.1   Effects of Intraventricular D_3 Antisense Infusion

Rats which received D_3 antisense S-oligo infusion (10 μg/μl × 1 μl/h) for 3 days did not show significant catalepsy (Figure 6), while their spontaneous locomotor activity was increased (Figure 7). These results suggest that D_3 receptors may have inhibitory effects on spontaneous locomotor activity. This is consistent with behavioral observation using the semi-selective D_3 receptor agonist 7-OH-DPAT, and antagonist U99194A. 7-OH-DPAT reduced locomotor activity in rats at doses that did not affect brain DA synthesis or release,[53,54] and U99194A was found to increase locomotion over a wide dose range.[55,56]

In addition, D_3 antisense treatment did not significantly influence d-amphetamine-induced locomotor activity (Figure 7), while it decreased quinpirole-induced locomotor activity and reduced d-amphetamine-induced stereotypy (Figures 8 and 9). Thus D_3 receptors may have a number of effects on the different regulation of motor responses.

4.2.2   Function of Striatal Postsynaptic D_3 Receptors

In contrast to the effects of D_2 antisense treatment, rats which had received unilateral intrastriatal multiple injections of D_3 antisense S-oligo (20 µg, twice a day for 3 days) did not show significant ipsilateral rotation to the DA agonists quinpirole or apomorphine (Figure 10). This suggests that D_3 receptors in the striatum have a different function than D_2 receptors, and that D_3 receptors do not mediate the motor activation induced by DA agonists.

4.2.3   Function of Somatodendritic and Terminal D_3 Receptors

Similar to the effects of D_2 antisense treatment, intranigral infusion of D_3 antisense also significantly attenuated the apomorphine-induced inhibition of the firing rate of SNC DA neurons. In addition, D_3 antisense treatment also significantly increased the proportion of striatal-evoked antidromic responses consisting of full spikes and decreased the antidromic threshold, indicating an increase in somatodendritic and terminal excitability, respectively. During unilateral intranigral D_3 antisense infusion, the rats also developed a contralateral rotation from the second day of infusion. Infusion of both D_2 and D_3 antisense oligos together seemed to produce an additive effect on these electrophysiological and behavioral responses.[20,21]

Pharmacological results also support a functional role for the D_3 receptor subtype in the autoreceptor-mediated regulation of DA neurons. It has been demonstrated that DA receptor agonist potencies for inhibition of SNC DA cell firing or DA synthesis correlate with dopamine D_3 receptor binding affinities.[57,58]

4.3   Effects of Antisense Knockout of D_4 Receptors

As with the D_3 receptors, D_4 receptors appear to be expressed at a lower level than D_2 receptors throughout the brain. The highest levels of D_4 receptor expression include the frontal cortex, midbrain, amygdala and medulla, with lower levels observed in the striatum and olfactory tubercle.[11,17] This distribution is largely overlapping with that of the D_2 and D_3 receptors although with much lower relative abundance.

After 3 days of i.c.v. infusion of D_4 antisense S-oligo, the rats did not show catalepsy (Figure 6), while their spontaneous locomotor activity was significantly decreased (Figure 7). Furthermore, the locomotor activity induced by quinpirole or d-amphetamine was also significantly decreased by D_4 antisense (Figures 7 and 8). Also the stereotypy induced by higher doses of d-amphetamine was significantly reduced by D_4 antisense (Figure 9).

Pharmacological studies have shown that one of the major differences between typical and atypical neuroleptics is that typical neuroleptics produce both reductions of locomotor activity and induce catalepsy, while atypical neuroleptics only produce reductions of locomotor activity without catalepsy.[59] D_4 receptor knockout produced no catalepsy, while decreasing locomotor activity, suggesting that D_4 receptors may mediate the antipsychotic effects of both typical and atypical neuroleptics. This is consistent with the observation that clozapine, the original atypical neuroleptic, has highest affinity for D_4 receptors.[41,60]

5. SUMMARY

While the cloning of five distinct DA receptor subtypes has produced a major challenge for the interpretation of DA-mediated biochemical and electrophysiological, as well as behavioral effects, molecular biology also provides us with the antisense strategy as an effective tool to approach these problems. In the study of DA receptors in the CNS, our laboratory and others have revealed that the antisense knockout strategy is both highly efficient and selective in reducing individual DA receptor subtypes selectively and that this receptor loss has observable and differential functional consequences. The efficacy of the strategy depends on several factors: the backbone modification of the oligodeoxynucleotide, the administration mode of the oligodeoxynucleotide, the stability of the target receptor protein, and the degree of target receptor loss required for a functional change. These experiments demonstrate the utility of this approach for the D2 receptor family, and hence, should be applicable for determining the behavioral functions of the D1 dopamine receptor subtypes as well. We further suggest that the application of antisense strategies may also be of general clinical utility for diseases where reducing CNS synaptic neurotransmission is therapeutic. For example, if the antipsychotic action of neuroleptic drugs is mediated by dopamine D_4 receptors, but their chronic use leads to a compensatory increase in D_2 receptor production and unwanted motor side effects such as tardive dyskinesia,[61] D_2 antisense treatment could overcome this problem by arresting D_2 receptor synthesis.

ACKNOWLEDGMENTS

This research was supported by MH-52383, MH-52450, a Johnson & Johnson Discovery Award and a Hoechst-Celanese Innovative Research Award.

REFERENCES

1. Carlsson A., Lindguist M., and Magnusson T., 3,4-Dihydroxy-phenylalamine and 5-hydroxytrytophan as reserpine antagonists, *Nature*, 180, 1200, 1957.
2. Dahlstrom A. and Fuxe K., Evidence for the existence of monoamine-containing neurons in the central nervous system. I. Demonstration of monoamines in the cell bodies of brain stem neurons, *Acta Physiol. Scand. (Suppl.)* 232, 1, 1964.
3. Hornykiewicz O., Dopamine and brain function. *Pharmacol. Res.*, 18, 925, 1966.
4. Marsden C. D., Dopamine and basal ganglia in human. *Semin. Neurosci.*, 4, 171, 1992.
5. Creese I., Burt D. R., and Snyder S. H., Dopamine receptor binding predicts clinical and pharmacological potencies of antischizophrenic drugs, *Science*, 192, 481, 1976.
6. Carlsson A., The current status of the dopamine hypothesis of schizophrenia, *Neuropsychopharmacology*, 1, 179, 1988.
7. Davis K. L., Kahn R. S., Ko G., and Davidson M., Dopamine in schizophrenia: a review and reconceptualization, *Am. J. Psychiatr.*, 148, 1474, 1991.
8. Kebabian J. W. and Calne D. B., Multiple receptors for dopamine, *Nature*, 277, 93, 1979.
9. Stoof J. C. and Kebabian J., Two dopamine receptors: Biochemistry, physiology, and pharmacology, *Life Sci.*, 35, 2281, 1984.
10. Hess E. J. and Creese I., Biochemical Characterization of Dopamine Receptors, in *Dopamine Receptors*, Creese I. and Fraser C.M., Eds., Alan R. Liss. Inc., New York, 1987, 1.
11. Sibley D. R. and Monsma F. J., Molecular biology of dopamine receptors, *Trends Pharmacol.*, 131, 61, 1992.
12. Civelli O., Bunzow J. R., and Grandy D. K., Molecular diversity of the dopamine receptors, *Annu. Rev. Pharmacol. Toxicol.*, 32, 281, 1993.
13. Zhou Q. -Y., Grandy D. K., Thambi L., Kushner J. A., Van Tol H. H. M., Cone R., Pribnow D., Salon J., Bunzow J. R., and Civelli O., Cloning and expression of human and rat D_1 dopamine receptors, *Nature*, 347, 76, 1991.
14. Grandy D. K., Zhang Y., Bouvier C., Zhou Q.- Y., Johnson R. A., Allen L., Buck K., Bunzow J. R., Salon J., and Civelli O., Multiple human D_5 dopamine receptor genes: a functional receptor and two pseudogenes, *Proc. Natl. Acad. Sci. U.S.A.*, 89, 9175, 1991.
15. Bunzow J. R., Van Tol H. H. M., Grandy D. K., Albert P., Salon J., Chisre M., Machida C. A., Neve K. A., and Civelli O., Cloning and expression of a rat D_2 dopamine receptor cDNA, *Nature*, 336, 783, 1988.
16. Sokoloff P., Giros B., Martres M.- P., Bouthenet M.- L., and Schwartz J.-C., Molecular cloning and characterization of a novel dopamine receptor (D_3) as a target for neuroleptics, *Nature*, 347, 146, 1990.
17. Van Tol H. H. M., Bunzow J. R., Guan H.-C., Sunahara R. K., Seeman P., Niznik H. B., and Civelli O., Cloning of a human dopamine D_4 receptor gene with high affinity for the antipsychotic clozapine, *Nature*, 350, 614, 1991.
18. Zhang M. and Creese I., Antisense oligodeoxynucleotide reduces brain dopamine D_2 receptors: behavioral correlates. *Neurosci. Lett.*, 161, 223-226, 1993.
19. Zhang M., Ouagazzal A., and Creese I., Differential roles for dopamine "D2" receptor subtypes in locomotor activity and stereotypy: antisense knockout study, *Soc. Neurosci. Abstr.*, 21, 364, 1995.
20. Martin L. P., Kita H., Sun B.-C., Zhang M., Creese I., and Tepper J. M., Electrophysiological consequences of D2 receptor antisense knockout in nigrostriatal neurons, *Soc. Neurosci. Abstr.*, 20:908, 1994.
21. Sun B.-C., Creese I., and Tepper J. M., Electrophysiology of antisense knockout of D2 and D3 dopamine receptors in nigrostriatal dopamine neurons, *Soc. Neurosci. Abstr.*, 21, 1661, 1995.

22. Tepper J.M., Sun B.-C., Martin L.P., and Creese I., Electrophysiological consequences of D_2 and/or D_3 receptor knockout by antisense oligodeoxynucleotides in nigrostriatal dopaminergic neurons. In: C. Ohye, M. Kimura and J. McKenzie (Eds.), *The Basal Ganglia V*, Plenum Press, New York, in press.

23. Baserga B. and Denhardt D. T. (Eds.), Antisense strategies. *Ann. N.Y. Acad. Sci.*, 660, 1992.

24. Cohen J., Oligonucleotides as therapeutic agents. *Pharmacol. Ther.*, 52, 211-225, 1991.

25. Murray J. A. H., *Antisense RNA and DNA*, John Wiley & Son, Inc., New York, 1993.

26. Wahlestedt C., Golanov E., Yamamoto S., Yee F., Ericson H., Yoo H., Inturrisi C. E., and Reis D. J., Antisense oligodeoxynucleotides to NMDA-R1 receptor channel protect cortical neurons from excitotoxicity and reduce focal ischaemic infarctions. *Nature*, 363, 260-263, 1993a.

27. Wahlestedt C., Pich M., Koob G. F., Yee F., and Heilig M., Modulation of anxiety and neuropeptide Y-Y1 receptors by antisense oligodeoxynucleotides. *Science*, 259, 528-531, 1993b.

28. Chiasson B. J., Hooper M. L., Murphy P. R., and Robertson H. A., Antisense oligonucleotide eliminates *in vivo* expression of c-*fos* in mammalian brain, *Eur. J. Pharmacol.*, 227, 451-453, 1992.

29. Loke S. L., Stein C. A., Zhang X. H., Mori K., Nakanishi M., Subasinghe C., Cohen J. S., and Neckers L. M., Characterization of oligonucleotide transport into living cells, *Proc. Natl. Acad. Sci. U.S.A.*, 86, 3474-3478, 1989.

30. Yakubov L. A., Deeva E. A., Zarytova V. F., Ivanova E. M., Ryte S., Yurchenk L. V., and Vlassov V. V., Mechanism of oligonucleotide uptake by cells: Involvement of specific receptors?, *Proc. Natl. Acad. Sci. U.S.A.*, 86, 6454-6458, 1989.

31. Wagner R. W., Gene inhibition using antisense oligodeoxynucleotides. *Nature*, 372, 333-335, 1994.

32. Campbell J. M., Bacon T. A., and Wickstrom E., Oligodeoxynucleotide phosphorothioate stability in subcellular extracts, culture media, sera and cerebrospinal fluid, *J. Biochem. Biophys. Methods*, 20, 259-269, 1990.

33. Agrawal S., Temsamani J., and Tang J. Y., Pharmacokinetics, biodistribution, and stability of oligodeoxynucleotide phosphorothioates in mice, *Proc. Natl. Acad. Sci. U.S.A.*, 88, 7595-7599, 1991.

34. Whitesell L., Geselowitz D., Chavany C., Fahmy B., Walbridge S., Alger J. R., and Neckers L. M., Stability, clearance, and disposition of intraventricularly administered oligodeoxynucleotides: implications for therapeutic application within the central nervous system, *Proc. Natl. Acad. Sci. U.S.A.*, 4665-4669, 1993.

35. Hall M. D., Jenner P., and Marsden C. D., Turnover of specific [³H]spiperone and [³H]N,n-propylnorapomorphine binding sites in rat striatum following phenoxybenzamine administration, *Biochem. Pharmacol.*, 32:2973-2977, 1983.

36. Norman A. B., Battaglia G., and Creese I., Differential recovery rates of rat D2 dopamine receptors as a function of aging and chronic reserpine treatment following irreversible modification: a key to receptor regulatory mechanisms, *J. Neurosci.*, 7, 1484-1491, 1987.

37. Leff S. E., Gariano R., and Creese I., Dopamine receptor turnover rates in rat striatum are age-dependent, *Proc. Natl. Acad. Sci. U.S.A.*, 81:3910-3914, 1984.

38. Hamblin M. and Creese I., Behavioral and radioligand binding evidence for irreversible dopamine receptor blockade by EEDQ, *Life Sci.*, 32, 2247-2255, 1983.

39. Meller E., Bordi F., and Bohmaker K., Behavioral recovery after irreversible inactivation of D-1 and D-2 dopamine receptors, *Life Sci.*, 44, 1019-1026, 1989.

40. Saller C. F., Kreamer L. D., Adamovage L. A., and Salama A. I., Dopamine receptor occupancy *in vivo*: measurement using N-ethoxycarbonyl-2ethoxy-1,2-dihydroquinoline (EEDQ), *Life Sci.*, 45, 917-929, 1989.

41. Seeman, P., Dopamine receptor sequences: therapeutic levels of neuroleptics occupy D_2 receptors, clozapine occupies D_4, *Neuropsychopharmacology*, 7, 261, 1992.
42. Yaswen P., Stampfer M., and Ghosh, K., Effects of sequence of thioated oligonucleotides on cultured human mammary epithelial cells, *Antisense Res. Dev.*, 3, 67-77, 1992.
43. Weiner D. M., Levey A. I., and Brann M. R., Expression of muscarinic acetylcholine and dopamine receptor mRNAs in rat basal ganglia, *Proc. Natl. Acad. Sci. U.S.A.*, 87, 7050-7054, 1990.
44. Creese I. and Iversen S. D., The pharmacological and anatomical substrates of the amphetamine response in the rat, *Brain Res.*, 83, 419–436, 1975.
45. Kelly P. H., Seviour P. W., and Iversen S. D., Amphetamine and apomophine responses in the rat following 6-OHDA lesions of the nucleus accumbens septi and corpus striatum, *Brain Res.*, 94, 507-522, 1975.
46. Worms P., Broekkamp C. L. E., and Lloyd K. G., Behavioral effects of neuroleptics. In: *Neuroleptics: Neurochemical, Behavioral, and Clinical Perspectives*. (Coyle J. T and Enna S. J. Eds.), New York: Raven Press, 1983, 93-118.
47. Kelley A. E., Lang C. G., and Gauthier A. M., Induction of oral stereotypy following amphetamine microinjection into a discrete subregion of the striatum, *Psychopharmacology*, 95, 556-559, 1988.
48. Weiss B., Zhou L.-W., Zhang S.-P., and Qin Z.-H., Antisense oligodeoxynucleotide inhibits D_2 dopamine receptor-mediated behavior and D_2 messenger RNA, *Neuroscience*, 55, 607-612, 1993.
49. Zhou L.-W., Zhang S.-P., Qin Z.-H., and Weiss B., *In vivo* administration of an oligodeoxynucleotide antisense to the D2 dopamine receptor messenger RNA inhibits D2 dopamine receptor-mediated behavioral and the expression of D2 dopamine receptors in mouse striatum, *J. Pharmacol. Exp. Ther.*, 268, 1015-1023, 1994.
50. Bunney B. S., Chiodo L. A., and Grace A. A., Midbrain dopamine system electrophysiological functioning: a review and new hypothesis, *Synapse*, 9, 79, 1991.
51. Tepper J. M. and Groves P. M., *In vivo* electrophysiology of central nervous system terminal autoreceptors, *Ann. N.Y. Acad. Sci.*, 604, 470, 1990.
52. Trent F. and Tepper J. M., Dorsal raphe stimulation modifies striatal-evoked antidromic invasion of nigral dopaminergic neurons *in vivo*, *Exp. Brain Res.*, 84, 620, 1991.
53. Svensson K., Carlsson A., Huff R. M., Kling-Petersen T., and Waters N., Behavioral and Neurochemical data suggest functional differences between dopamine D2 and D3 receptors, *Eur. J. Pharmacol.*, 263, 235-243, 1994a.
54. Svensson K., Carlsson A., and Waters N., Locomotor inhibition by the D3 ligand R-(+)-7-OH-DPAT is independent of changes in dopamine release, *J. Neural Transm.*, 95, 71-74, 1994b.
55. Waters N., Lofberg L., Haadsma-Svensson S., R. Svensson K., Sonesson C., and Carlsson A., Differential effects of dopamine D2 and D3 receptor antagonists in regard to dopamine release, *in vivo* receptor displacement and behaviour, *J. Neural. Transm.*, 98, 39-55, 1994.
56. Waters N., Svensson K., Haadsma-Svensson S. R., Smith M. W., and Carlsson A., The dopamine D3-receptor: a postsynaptic receptor inhibitory on rat locomotor activity, *J. Neural Transm.*, 94, 11-19, 1993.
57. Kreiss D. S., Bergstrom D. A., Gonzalez A. M., Huang K.-X., Sibley D. R., and Walters J. R., Dopamine receptor agonist potencies for inhibition of cell firing correlate with dopamine D_3 receptor binding affinities, *Eur. J. Pharmacol.*, 277, 209, 1995.
58. Meller E., Bohmaker K., Goldstein M., and Basham D. A., Evidence that striatal synthesis-inhibitory autoreceptor are dopamine D3 receptors, *Eur. J. Pharmacol.*, 249, R5, 1993.

59. Reynolds G. P., Developments in the drug treatment of schizophrenia, *Trends Pharmacol. Sci.*, 13, 116, 1992.
60. Seeman P., Guan H.-C., and Van Tol, H. H. M., Dopamine D4 receptors elevated in schizophrenia, *Nature*, 365, 441, 1993.
61. Burt D. R., Creese I., and Snyder, S. H., Antischizophrenic drugs: chronic treatment elevates dopamine receptor binding in brain, *Science*, 196, 326, 1977.

Part II

NEUROTRANSMITTER INTERACTIONS REGULATING STRIATAL GENE EXPRESSION

Chapter **4**

GLUTAMATERGIC AND CHOLINERGIC-REGULATION OF IMMEDIATE-EARLY GENE AND NEUROPEPTIDE GENE EXPRESSION IN THE STRIATUM

John Q. Wang and Jacqueline F. McGinty

CONTENTS

0-8493-8550-4/96/$0.00+$.50
© 1996 by CRC Press, Inc.

1. INTRODUCTION

In recent years, the striatum has been the focus of studies on brain mechanisms underlying the behavioral effects of drugs of abuse. The psychostimulants, cocaine and amphetamine, initiate changes in behavior by increasing the extracellular concentration of dopamine released from mesolimbic and mesostriatal systems. Dopamine D_1 receptor stimulation activates gene expression in striatal medium-sized spiny neurons which project to the substantia nigra. Among inducible genes, the immediate early genes (IEG), c-*fos* and *zif* 268, and the neuropeptide genes, preprodynorphin (PPD) and substance P (SP), have been studied most extensively. Changes in the expression of these genes are thought to constitute adaptive changes in cellular physiology and contribute to the molecular plasticity which underlies long-term changes in behavior induced by drugs of abuse.

Dopamine D_1 receptors are the best characterized mediators of striatal gene expression induced by psychotropic drugs (see Chapters 1 and 5). D_1 receptors are unequivocally expressed by most, if not all, striatonigral neurons[1-3] and D_1 receptor antagonists block basal and psychostimulant-induced gene expression in these neurons.[4-7] However, despite the controversy surrounding whether or not D_2 receptors are expressed by striatonigral neurons,[8] D_2 receptor antagonists also block psychostimulant-induced, though not basal levels, of dynorphin immunoreactivity in the striatum.[6,7] More recently, D_2 antagonists have been demonstrated to significantly attenuate psychostimulant-induced

IEG and PPD mRNA levels.[9-11] Therefore, D_1/D_2 synergy is a phenomenon which underlies changes in gene expression (see Chapter 1) as well as acute electrophysiological events.

In addition to the effects that dopamine receptor stimulation has on striatonigral PPD and SP gene expression, some publications report that psychostimulants slightly, but significantly, increase preproenkephalin (PPE) mRNA levels in the dorsal striatum,[10,12-14] an effect which is mediated by D_1, but not D_2, receptors.[10] PPE mRNA is widely thought to be confined to striatopallidal neurons in which it is induced strongly by dopaminergic denervation or D_2 receptor blockade.[15-18] However, D_2 agonists have little, if any, effect on PPE mRNA or enkephalin levels in the striatum,[19-20] indicating that endogenous dopamine is saturating D_2 receptors under basal conditions.

In addition to the strong regulation of gene expression in the medium spiny neurons by dopamine receptors, recent studies indicate that glutamatergic and cholinergic neurotransmission strongly modulate the effects of dopamine on striatal gene expression. Excitatory amino acid (EAA) receptor antagonists and muscarinic receptor agonists both block dopamine agonist-induced behaviors and striatonigral gene expression. Therefore, acetylcholine and glutamate must interact cooperatively with dopamine to mediate the actions of these psychostimulants in the striatum as well as in related extra-striatal sites. Emerging studies on the detailed intracellular cascades triggered by these neurotransmitter interactions (see Chapter 5) will provide further insight into the molecular pathophysiology of psychostimulant effects.

This chapter reviews first the role of glutamatergic transmission in the regulation of behavioral, and particularly genomic, responses in dorsal striatal neurons to dopaminergic stimulation. The possible mechanisms of EAA/dopamine interactions are then discussed in detail. In the second part of this chapter, we review the cholinergic regulation of striatal gene expression under normal and dopamine-stimulated conditions in intact animals and the possible mechanisms underlying the opposite effects of muscarinic agents on striatonigral vs. striatopallidal neurons. Finally, an overall view of dopamine-cholinergic-EAA interactions, which extends the model first proposed by Di Chiara and colleagues,[20a] will conclude this chapter.

2. GLUTAMATERGIC REGULATION OF STRIATAL GENE EXPRESSION

2.1   Organization of the Glutamatergic System in the Striatum

Numerous studies have demonstrated abundant projections to the striatum from multiple forebrain regions which use glutamate as a transmitter. Briefly, dorsal striatum (caudatoputamen) receives glutamatergic inputs from widespread areas of the cerebral cortex and thalamus whereas ventral striatum (nucleus accumbens) receives glutamatergic projections from the amygdala, thalamus, ventral subiculum, and prefrontal cortex. Glutamatergic terminals make asymmetrical (excitatory) synaptic contacts with striatal neurons, including medium-sized spiny projection neurons[21-24] and aspiny interneurons (GABAergic or cholinergic neurons).[25,26] As the medium spiny neurons are also major synaptic targets of dopaminergic terminals,[27,28] it is plausible that the activity of a single striatal projection neuron can be modulated postsynaptically by both glutamate and dopamine inputs. This notion is supported by direct convergence of glutamatergic and dopaminergic terminals on the dendritic spines of the same striatal output cells.[23,29,30] However, the paucity of axoaxonic synapses within the striatum[29] and a lack of direct synaptic contacts between the two major incoming terminals[23,27,29,31,32] restricts the possibility of classical synaptic contacts between glutamate and dopamine terminals. Alternatively, presynaptic glutamate/dopamine interactions may be mediated by nonsynaptic contacts between terminals[23,31] that express extrasynaptic heteroreceptors and are activated or inhibited by diffusion of neurotransmitter away from the synapse where it is released.

High binding levels of all three pharmacologically identified subtypes of ionotropic EAA receptors, N-methyl-D-aspartate (NMDA), kainate and AMPA,[33] are found in the striatum.[34-37] However, detailed cellular localization of EAA receptors in the striatum needs to be determined in order to understand the precise mechanisms underlying glutamate/dopamine interactions. The possibilities include (1) pre- and postsynaptic localization of receptors, (2) localization of receptors in specific subpopulation of striatal neurons, and (3) colocalization of EAA and dopamine receptors on the dendrites of the same neurons. The majority of affinity binding sites in the dorsal striatum for all three subtypes of ionotropic receptors appears to be localized postsynaptically; approximately 92% NMDA, 81% kainate, and 80% AMPA receptors have been found on postsynaptic neurons,[38] although it has been suggested that 10% of EAA receptors are presynaptically localized in the striatum.[39,40] Consistent with a predominant postsynaptic receptor distribution, 6-hydroxydopamine (6-OHDA)-induced lesions of the nigrostriatal neurons or decortication have little or no effect on the density of striatal EAA receptors.[38,39,41,42] Binding sites of all three types of EAA receptors are present on striatonigral projection neurons.[43] Both EAA and dopamine receptors may reside on the same striatal neurons since toxic EAA receptor agonists cause parallel reduction of binding levels of EAA (all three subtypes) and D_1 dopamine receptors.[44,45]

Recent exciting advances in molecular EAA receptor research have demonstrated multiple subunits for each ionotropic receptor.[46-49] The

NMDA receptor is composed of two families of subunits in the rat, NMDAR1 encoded by a single gene but alternatively spliced to produce eight distinct isoforms (NMDAR1A-H), and NMDAR2 encoded by four separate genes (NMDAR2A-D). Nine different subunits for non-NMDA receptors have been cloned from rat brain, GluR1-4 (also known as GluR-A—GluR-D) for AMPA and GluR5-7, KA1, and KA2 for kainate receptors. Each subunit appears to have its own pharmacological characteristics and distinct pharmacological effects occur with activation of different subunit combinations. Subunit cloning has provided an opportunity to examine the cellular and subcellular distribution of receptor subunits in brain by *in situ* hybridization and immunocytochemistry. To date, mRNAs encoding NMDAR1, NMDAR2A-D, and GluR1-7 have been demonstrated in rat striatum with different subunit-specific densities.[50-54] Newly developed AMPA subunit-specific antisera have been utilized in order to delineate the localizations of four AMPA subunits on subpopulations of striatal neurons in immunocytochemical studies.[45,55,56] In a double labeling study, cholinergic neurons were demonstrated to contain GluR1 and GluR4 immunoreactivity whereas medium spiny neurons contained GluR1 and GluR2/3/4 immunoreactivity.[55]

Metabotropic glutamate receptors (mGluR) are members of a novel family of EAA receptors which are coupled to multiple second messenger systems via G proteins.[46,48,57,58] Seven subunits of mGluR (mGluR1-7) have been cloned and, like ionotropic receptors, they are heterogeneous in their distribution, pharmacology, and coupling with intracellular effectors. Studies on *Xenopus* oocytes expressing the cloned receptors have linked (1) mGluR1 and 5 to increases in phosphoinositide hydrolysis, (2) mGluR1 to adenylate cyclase-independent increases in cAMP formation, (3) mGluR2-4, 6-7 to adenylate cyclase-dependent decreases in cAMP formation, and (4) several mGluRs to modulation of ion channel activity.

Quantitative receptor autoradiography has revealed high levels of mGluR binding sites in the striatum.[34] Like ionotropic receptors, 90% of mGluRs are located on postsynaptic striatal neurons.[40] Studies with *in situ* hybridization[59-64] and immunocytochemistry[65-68] demonstrate the presence of mGluRs 1-5 and 7 in the rat striatum with high levels of mGluR3 and 5 and moderate levels of mGluR1, 4, and 7. MGluR2 is expressed only in a small population of large polygonal neurons, probably cholinergic interneurons.[63]

2.2. Glutamatergic Regulation of Behavioral Responses to Psychostimulant Administration

NMDA and kainate/AMPA receptors mediate acute and long-term behavioral alterations in response to psychostimulant administration.

With pharmacological blockade of NMDA receptors by MK-801, Karler and colleagues[69,70] first reported prevention of initiation of behavioral sensitization in response to repeated amphetamine. Similar findings have been reported by other laboratories.[71-73] Furthermore, the role of NMDA receptors in the development of sensitization has been extended to repeated methamphetamine[74] and cocaine.[75] Karler and colleagues also demonstrated that the kainate/AMPA antagonist, 6,7-dinitro-quinoxaline-2,3-dione (DNQX), prevents initiation and expression of amphetamine-induced behavioral sensitization and initiation of cocaine-induced behavioral sensitization.[75,76] With regard to glutamatergic participation in the acute actions of psychostimulants, systemic NMDA and non-NMDA antagonists, when administered at a relatively high dose in mice, block the acute motor stimulating effects of amphetamine, methamphetamine, or cocaine.[77-79] The locus of this blocking action appears to be in the striatum because microinjection of two different EAA receptor antagonists into dorsal striatum or nucleus accumbens blocks systemic amphetamine- or cocaine-induced stereotypies, hypermotility, and conditioned place preference, respectively.[78,80-84] Furthermore, enhancement of local glutamatergic tone by intrastriatal EAA receptor agonists, such as NMDA, quisqualate, or kainate, induces behavioral changes similar to those induced by stimulants.[85-88]

The functional role of striatal mGluRs in motor control was demonstrated recently in a report that a unilateral intrastriatal injection of the mGluR agonist, 1S,3R-ACPD, causes contralateral turning[89,90] and seizures.[91] These behavioral changes are not blocked by MK-801 and DNQX but are blocked by the systemic mGluR antagonist, AP3, and by dantrolene, an inhibitor of intracellular Ca^{2+} mobilization, indicating that phosphoinositide-linked mGluRs in the rat striatum participate in extrapyramidal motor control.[91]

2.3 Glutamatergic Regulation of IEG and Opioid Peptide Gene Expression in the Striatum

Glutamatergic regulation of IEG and neuropeptide expression in the striatum may be an essential element of the neuronal adaptation which underlies the delayed or long-term effects of psychotropic agents.[92] Johnson et al.[93,94] and Singh et al.[95] first demonstrated that NMDA receptors govern the upregulation of striatal dynorphin immunoreactivity induced by cocaine, methamphetamine, or 3,4-methylenedioxymethamphetamine. Stimulant-induced Fos protein induction in striatal neurons has been tied also to NMDA receptor regulation.[96-99] Furthermore, transection of corticostriatal afferents reduces,[100] and cortical stimulation increases,[101] Fos expression in striatal neurons. Recently, with *in situ* hybridization, we found that a single injection of amphetamine or methamphetamine is capable of stimulating mRNA expression of the IEGs, c-*fos*, and *zif* 268, and prolonged induction of opioid peptide mRNA in the striatum.[102-105] The NMDA receptor antagonists, MK-801 and CPP, and the kainate/AMPA receptor antagonist, DNQX, attenuated the increases in striatal *zif* 268, PPD, and PPE expression induced by acute amphetamine.[106-107] Similar results were obtained subsequently with acute methamphetamine except that strong induction of *zif* 268 was still present after pretreatment with DNQX.[108] These data indicate that NMDA and kainate/AMPA receptors mediate amphetamine-stimulated IEG and opioid peptide expression in striatal neurons.

The above studies demonstrated that the mRNA blocking effects of NMDA and non-NMDA antagonists occurred at doses which did not affect the acute behavioral effects of amphetamines. Therefore, EAA receptors may preferentially modulate those cellular responses which lead to development of long-term effects of dopamine receptor stimulation. Synaptic activation of dopamine receptors leads to multiple biological responses in postsynaptic neurons via different intracellular messengers. Included are the relatively short-lived release of neurotransmitters that mediates the acute behavioral effects of stimulants and the relatively long-lived changes in gene expression which underlie neuronal plasticity indicative of drug sensitization or addiction. If these long-lived biochemical events are more susceptible to EAA receptor blockade than short-lived neurotransmitter release, it would follow that EAA receptor blockade may readily prevent changes in gene expression and leave acute behavioral effects of stimulants unaffected. This theory agrees with the observation that both NMDA and kainate/AMPA antagonists blocked behavioral sensitization to repeated amphetamine, but they were less effective in altering behaviors induced by acute injection of this drug.[69,70,76]

Although not yet documented experimentally, it is likely that the mRNA blocking effects of systemically administered EAA receptor antagonists take place in the striatum because intrastriatal antagonists mimic the blocking effects of peripheral antagonists on stimulant-induced behaviors.[78,80-84] In addition, direct application of non-toxic doses of NMDA agonists into the striatum *in vivo* or *in vitro* induces expression of c-*fos* and/or *zif* 268[109-111] and PPE[112] mRNAs.

Both NMDA and kainate/AMPA receptor antagonists substantially reduce basal levels of *zif* 268 expression in the striatum.[106,108] Stable and relatively high basal levels of *zif* 268 are expressed in the rat brain, including striatum and cortex, in contrast to its transient induction after dopamine stimulation. The magnitude of basal *zif* 268 expression in the brain is higher than that of c-*fos*, the expression of which is negligible in normal animals. The high basal level of *zif* 268 expression provides a unique opportunity to identify regulators which are negatively coupled to the expression of this gene. The decrease in basal striatal *zif*268 after pharmacological blockade of two different EAA receptor subtypes indicates that tonic glutamatergic activity is needed for basal activity of *zif*268 gene expression in the striatum. In contrast to *zif*268, constitutive expression of PPD and PPE mRNA appears to be more resistant to NMDA and kainate/AMPA receptor blockade. Basal opioid peptide mRNA expression in the striatum does not change 3.5 h after MK-801, CPP, or DNQX.[106-108] However, since other time points were not examined, it is possible that the effect of EAA receptor antagonists on the basal levels of opioid peptide gene expression is either a dose- and time-related event or may need simultaneous blockade of both NMDA and kainate/AMPA receptors. Indeed, Somers and Beckstead[113] found that 8 h, but not 4 h, after infusion of CPP into the lateral ventricle, there was a significant decrease in PPE mRNA levels in the dorsal striatum. Furthermore, destruction of corticostriatal afferents, which presumably eliminated both NMDA and kainate/AMPA tone, also decreased the basal level of PPE mRNA expression in the striatum.[114,115]

In addition to ionotropic EAA receptors, activation of mGluRs is needed for mRNA induction of the IEGs, c-*fos*, *junB*, and *zif*268, in primary striatal cultures from striatum.[111] In addition, we recently found that intrastriatal injection of the selective mGluR receptor antagonist, (±)-α-methyl-4-carboxyphenylglycine, attenuated neuropeptide expression induced by systemic amphetamine.[116]

Pharmacological blockade of D_2 receptors or striatal monoamine depletion by 6-OHDA or reserpine is well known to induce IEG and PPE gene expression in striatopallidal neurons[117] (see also Chapters 2, 6, and 7). The gene induction occurs as a direct result of decreased dopaminergic tone on these neurons. However, NMDA antagonists attenuate Fos immunoreactivity induced by haloperidol[118] or dopamine depletion.[119] Therefore, Fos induction by neuroleptics or dopaminergic depletion is also regulated by NMDA receptors.

2.4   Mechanisms Underlying Glutamate and Dopamine Interactions

Although the exact mechanisms of glutamate/dopamine interactions in regulation of gene expression in striatal neurons in response to drug stimulation are far from clear, the following possibilities should be taken into consideration. First, tonic glutamate activity may serve to maintain the excitability of dopaminergic neurons and striatal neurons. Blockade of this excitatory synaptic activity may result in a reduction of responsiveness of those neurons to incoming stimuli. Because EAA receptors are tonically active during normal excitatory transmission in the cerebral cortex[120,121] and striatum[122] and tonic glutamate activity is necessary for constitutive expression of zif 268[123] and PPE[113-115] mRNA in the striatum, the ability of NMDA and non-NMDA antagonists to block stimulant-induced behaviors and gene expression in striatal neurons may not be readily distinguished from a general reduction of activity. An overall reduction of spontaneous behaviors and normal activity of zif 268 expression in cortical and striatal regions by CPP and DNQX alone supports this concept.[106-108] However, CPP and DNQX blocked amphetamine- or methamphetamine-stimulated PPD and PPE expression without affecting the basal level of gene expression in our experiments.[106-108] This finding argues against the possibility that CPP and DNQX blocked stimulated opioid gene expression through a general reduction in neuronal activity. Instead, it is more likely that, by blocking postsynaptic EAA receptors, the effects of postsynaptic dopamine receptor stimulation were disenabled; i.e., synergistic activation of both receptor systems is necessary to stimulate gene expression in medium spiny neurons.

The second consideration involves the mechanism(s) underlying the interactions between the glutamate and dopamine systems which regulate striatal gene expression in response to stimulant administration. The major question is: To what extent do presynaptic vs. postsynaptic events control these interactions?[124] There is evidence that D_2 receptor agonists inhibit glutamate release by stimulating D_2 heteroceptors on corticostriatal terminals.[125,126] There is also evidence that systemic infusion of cocaine or local infusion of the D_1 receptor agonist, SKF-82958, evokes glutamate release in the ventral tegmental area (VTA),[127] suggesting that D_1 receptors may reside on glutamatergic terminals in the ventral midbrain. Furthermore, there is some evidence that dopamine receptor stimulation elicits glutamate release in the striatum. Local infusion of apomorphine evokes increased extracellular glutamate levels in the striatum.[128,129] Moreover, amphetamine[130] and cocaine[131] increase extracellular glutamate levels 30–40 min after injection in the neostriatum and nucleus accumbens, respectively. In addition, methamphetamine or amphetamine plus iprindole (which decreases the breakdown of amphetamine) evokes a delayed increase in glutamate release in the neostriatum.[132,133] The most likely sequence of events is that dopamine released by amphetamines or cocaine stimulates increased glutamate release via a polysynaptic mechanism. Karler and co-workers[81] provided evidence that intrastriatal infusions of D_2, NMDA, and $GABA_A$ agonists all induce stereotypies and that antagonists of all three systems block the stereotypical effects of cocaine or amphetamine in mice. However, the $GABA_A$ antagonist, bicuculline, but not the D_2 antagonist, sulpiride, blocked the stereotypical effects of intrastriatal NMDA. Furthermore, neither sulpiride or the NMDA antagonist, CPP, blocked the stereotypical effects of the GABA agonist, THIP, administered intrastriatally. These data indicate that the sequence of events underlying the stereotypical behavior induced by stimulants is initiated by dopaminergic activation, followed by increased glutamatergic activity which together stimulate GABAergic outflow from the striatum (see Figure 1). This study proposes an important model of striatal glutamate and dopamine interactions to explain the marked effects of EAA receptor antagonists on behaviors stimulated by acute and repeated administration of dopamine agonists. The synergistic dopamine-glutamate activity is in accordance with increased electrophysiological activity of motor-related cells in the neostriatum evoked by amphetamine.[134] However, these data still do not tell us precisely how dopamine increases glutamatergic neurotransmission, only that effects are, indeed, occurring in the striatum and are likely to be due to a polysynaptic mechanism.

With regard to glutamatergic modulation of dopamine release, there are many reports that glutamate agonists increase dopamine release with the interpretation that a presynaptic facilitatory effect of glutamate receptors on dopaminergic terminals is responsible.[135-139] However, most of these studies have demonstrated that high (mM) concentrations of glutamate or selective glutamate receptor agonists are necessary to influence dopamine release in the striatum. At these concentrations, Moghaddam and colleagues[137] demonstrated that extracellular K^+ concentrations were massively elevated and signs of spreading depression were detected. These data do not support a selective facilitatory effect of glutamate on presynaptic dopamine release under physiological conditions. However, at least three groups have reported that systemic or intrastriatal administration of glutamate receptor antagonists significantly attenuates psychostimulant-induced, but not basal, dopamine levels.[140-142] If presynaptic facilitatory glutamate receptors on dopaminergic terminals are not responsible, as suggested by Keefe and colleagues,[142a] then EAA antagonists must somehow decrease dopamine cell firing in the ventral mesencephalon. Because intrastriatal administration was as effective as systemic administration of EAA antagonists, a decrease in dopamine neuronal firing rate in the substantia nigra would require an increase in striatonigral GABAergic neuronal activity evoked by glutamate receptor blockade. This counterintuitive result could only occur if the predominant effect of glutamate receptor blockade was to turn off a tonically active inhibitory interneuron that impinges on striatonigral cells. The most likely candidates for this job are the GABA/parvalbumin-containing interneurons which receive glutamatergic inputs and project to cholinergic and medium spiny neurons in the striatum.[143]

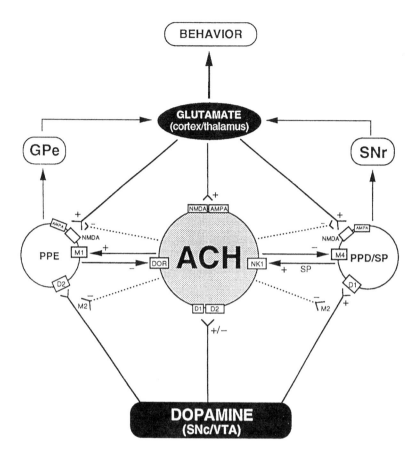

FIGURE 1.

Schematic illustration of a proposed model of the acetylcholine/dopamine/glutamate interactions that regulate striatal gene expression. Glutamatergic inputs from the cerebral cortex and thalamus stimulate medium spiny neurons containing preproenkephalin (PPE) or preprodynorphin (PPD)/substance P (SP) and cholinergic interneurons (ACH) in the striatum via NMDA and AMPA receptors. Cholinergic neurons inhibit gene expression in striatonigral neurons through M_4 receptors and stimulate that of striatopallidal neurons through M_1 receptors. Dopaminergic inputs from substantia nigra pars compacta (SNc) and ventral tegmental area (VTA) stimulate gene expression in striatonigral neurons through D_1 receptors and inhibit that of striatopallidal neurons through D_2 receptors. Dopaminergic transmission also exerts a D_1-dependent facilitatory influence and a D_2-dependent inhibitory influence on acetylcholine release. Local collaterals of striatonigral neurons augment acetylcholine release via NK_1 receptor stimulation and local collaterals of striatopallidal neurons reduce acetylcholine release via delta opioid receptors (DOR). In addition, stimulation of muscarinic receptors may presynaptically inhibit dopamine or glutamate release as indicated by dotted lines. Further discussion of the functional roles of these interactions in regulation of immediate early gene, PPE, PPD, and SP gene expression can be found in the text. Extra-striatal mechanisms, not illustrated here, also contribute to the regulation of medium spiny neuronal gene expression under normal conditions and in response to psychostimulant administration. Abbreviations also include GPe, globus pallidus external segment and SNr, substantia nigra pars reticulata. (Modified from Di Chiara, G., Morelli, M., and Consolo, S., *Trends Neurosci.*, 17, 288, 1994. With permission.)

2.5   Summary

Glutamatergic transmission mediates long-term behavioral alterations and striatal gene expression induced by indirect dopamine agonists. Both NMDA and kainate/AMPA subtypes of EAA receptors are implicated in this behavioral and genomic regulation. Although the precise mechanisms of glutamate/dopamine interactions are unknown, functional studies suggest that both presynaptic and postsynaptic interactions between the two systems regulate medium spiny neuronal gene expression. In addition to the functional role of all classes of ionotropic EAA receptors, involvement of the mGluR system also is implicated.

3. CHOLINERGIC REGULATION OF STRIATAL GENE EXPRESSION

3.1   Organization of the Cholinergic System in the Striatum

The striatum contains a population of cholinergic neurons which can be revealed by acetylcholinesterase (AChE) histochemistry or choline acetyltransferase (ChAT) immunohistochemistry.[144] These cholinergic neurons are interneurons and have large cell bodies with aspiny dendrites which account for about 1–2% of the striatal neuronal population.[145,146] Cholinergic neurons are distributed throughout the striatum in a widely dispersed pattern but they may also have regionally specific influences, because they are more prominent in the lateral caudate-putamen[147,148] and in the middle portion of the rostrocaudal axis of the nucleus accumbens.[149] Despite their small number, the highly branched axons of cholinergic neurons reach all types of surrounding neurons in the entire striatum, with particularly dense innervation of medium spiny projection neurons.[150]

Striatal cholinergic neurons receive synaptic inputs from two major extrinsic sources: mesencephalic dopaminergic neurons[151,152] and glutamatergic neurons arising from the thalamus or, less frequently, from the cerebral cortex.[153,154] Most cholinergic neurons express D_2 receptors,[155,156] EAA ionotropic receptors,[53] and mGluR2 receptors,[63] but only 30% of striatal cholinergic neurons express D_1 receptors.[157] In addition to extrinsic inputs, cholinergic neurons also receive synaptic input from intrinsic medium spiny neurons,[158] i.e., striatonigral and striatopallidal neurons. Recent *in situ* hybridization studies have identified striatal cholinergic neurons as a primary neuronal population expressing neurokinin NK_1 receptors,[155,159-161] the physiological receptor for SP, and delta opioid receptors,[162,163] the physiological receptor for enkephalin. These two latter receptors mediate selective signaling from medium spiny neurons containing either SP or enkephalin which contact cholinergic neurons.

Cholinergic neurons innervate medium spiny neurons[150,164] and medium aspiny interneurons containing somatostatin/neuropeptide Y, NO synthetase, and GABA. The aspiny neurons also receive glutamatergic inputs and project to cholinergic neurons and medium spiny neurons.[143,165,166] Furthermore, there are appositions between dopaminergic and cholinergic terminals which suggest presynaptic interactions.[164] Unfortunately, because they are so scarce, there are few detailed studies of the electrophysiological consequences of cholinergic cellular activity. However, Wilson and colleagues characterized a small set of these neurons by their tonic, random firing and how readily they fired action potentials in response to stimulation by relatively few afferents.[167] They concluded that the large, aspiny neurons with their extensive axonal and dendritic fields were in a strategic position to regulate the excitability of medium spiny neurons. In addition, muscarinic receptor stimulation appears to have a predominant effect on release of presynaptic neurotransmitters, including dopamine, via reduction of Ca^{2+} influx and a secondary impact on slow EPSPs of frequently encountered striatal neurons.[168]

In situ hybridization and immunohistochemistry have demonstrated the cellular distribution of M_1-M_4 subtypes of muscarinic receptors in the rat striatum.[169,170] Most cholinergic neurons (80% or more) express M_1, M_2, and M_4 receptors. Almost all the SP/PPD neurons contain M_1 and M_4 receptors. All enkephalinergic neurons contain M_1 receptors, but only 39% contain M_4 receptors.[169] Of these four subtypes, M_1 and M_3 receptors are coupled to the activation of phosphoinositide hydrolysis whereas M_2 and M_4 are coupled to inhibition of adenylate cyclase.[171,172]

3.2   Muscarinic Regulation of Behavioral Responses to Psychostimulant Administration

The role of the striatal cholinergic system in movement control has been a major topic of research for decades. Early behavioral studies demonstrated that intrinsic acetylcholine activity generally opposed the actions of dopamine in a dynamic acetylcholine/dopamine balance that controls basal ganglia function. For example, dopamine receptor agonists, primarily through D_1 activation, increase, and muscarinic receptor agonists decrease, spontaneous locomotor activity.[173,174] D_2 dopamine receptor antagonists induce catalepsy whereas muscarinic receptor antagonists increase behavioral activity.[175,176] Furthermore, pharmacological blockade of muscarinic receptors enhances the motor stimulating effects of the indirect dopamine agonists, amphetamine and cocaine,[173-176] and the direct agonist, apomorphine,[177] and attenuates catalepsy induced by D_2 antagonists.[177] In contrast, muscarinic receptor activation attenuates amphetamine-stimulated motor activity.[173,174,176] Additionally, the effects of chronic treatment with scopolamine on behavioral sensitization to methamphetamine have been reported in two recent articles. In one study, when given at a high dose (3 mg/kg), repeated scopolamine blocked behavioral sensitization evoked by repeated methamphetamine.[178] In contrast, when given at a low dose (0.5 mg/kg), scopolamine potentiated repeated methamphetamine-induced behavioral sensitization.[179] The precise mechanism for these paradoxical findings is unclear and the actual role played by striatal acetylcholine in the development and expression of behavioral sensitization remains to be determined.

3.3   Regulation of Acetylcholine Release by Dopaminergic and Glutamatergic Inputs

Recent *in vivo* microdialysis studies have demonstrated a reciprocal regulation of striatal acetylcholine release by D_1 and D_2 receptors. Pharmacological activation of D_1 receptors increases, whereas D_1 receptor blockade decreases, acetylcholine release.[180-183] In contrast, D_2 receptor activation decreases, whereas D_2 receptor blockade increases, acetylcholine release.[180,183-185] Thus, dopamine controls acetylcholine release in a facilitatory manner via stimulation of D_1 receptors and in an inhibitory manner via stimulation of D_2 receptors. This D_1/D_2 reciprocal control of acetylcholine release appears to be balanced under basal conditions since simultaneous blockade of both receptors or acute dopamine synthesis inhibition does not affect basal levels of extracellular acetylcholine.[186,187] Interestingly, co-activation of D_1 and D_2 receptors by systemic administration of indirectly acting dopamine agonists results in a net increase in extracellular acetylcholine levels as measured by *in vivo* microdialysis.[184,185,188,189] However, recent evidence indicates that the level of AChE inhibition in the perfusate influences this result.[190,191]

The actions of D_1 receptors on acetylcholine release appear to be indirect whereas those of D_2 receptors are direct. However, there is disagreement over whether intrastriatal or extrastriatal D_1 receptors mediate basal and stimulated acetylcholine release. Intrastriatal administration of D_1 antagonists is reported to decrease,[185] or not affect,[184,192] basal and dopamine agonist-stimulated acetylcholine release. In addition, intranigral infusion of SCH-23390 blocked amphetamine-induced acetylcholine release in the striatum.[192] Both intranigral and intrastriatal D_1 receptors are probably important and both exert their influence on striatal acetylcholine release in an indirect, polysynaptic manner. In the striatum, dopamine release stimulates SP neurons via D_1 receptors and results in local release of SP[193] which, in turn, activates cholinergic neurons via NK_1 receptors.[194-197] In the ventral mesencephalon, D_1 receptors are located on GABAergic terminals and facilitate GABA release in the VTA[198] and in the substantia nigra (Yamamoto, B.K., personal communication). Increased GABA release would inhibit nigrothalamic GABA neurons, disinhibiting thalamostriatal and thalamocortical/corticostriatal glutamate release. Indeed, there is evidence that intrastriatal infusion of MK-801 blocks the parafascicular thalamic-stimulated increase in acetylcholine release in the presence of a lesion of corticostriatal afferents.[199] In addition, the corticostriatal lesion itself reduced basal levels of acetylcholine. Finally, intrastriatal infusion of NMDA or AMPA increases acetylcholine release and local infusion of either NMDA[184,200-202] or kainate/AMPA receptor antagonists[192] block basal and amphetamine-induced acetylcholine release in the striatum.

3.4   Effects of Muscarinic Cholinergic and D_1 Dopaminergic Inputs Converging on Striatonigral Neurons

Under normal conditions, tonic dopaminergic stimulation of D_1 receptors maintains IEG and neuropeptide gene (PPD/SP) expression in striatonigral neurons. Blockade of D_1 receptors reduces basal gene expression in these neurons. In contrast to this facilitatory D_1 action, acetylcholine tonically inhibits striatonigral gene expression. For example, muscarinic receptor blockade significantly increases Fos immunostaining predominantly in PPD/SP-containing neurons.[203] With *in situ* hybridization, we recently found that the nonselective muscarinic receptor antagonist, scopolamine, upregulated PPD and SP mRNA expression in the dorsal and ventral striatum in a dose-dependent fashion.[204] However, exogenous administration of the muscarinic receptor agonist, oxotremorine, did not alter constitutive expression of PPD and SP mRNAs.[204] These data indicate that inhibitory regulation by endogenous acetylcholine appears to function at its maximal level under normal conditions. Acetylcholine also diminishes dopamine-stimulated gene expression in striatonigral neurons. Blockade of muscarinic receptors substantially potentiated the effects of amphetamine or the full D_1 receptor agonist, SKF-82958, on IEG (c-*fos* and *zif*268) and peptide gene (PPD and SP) expression in intact rats.[205,206] Conversely, oxotremorine attenuated amphetamine-stimulated PPD and SP gene expression.[204] In addition, acute scopolamine administration potentiated D_1 agonist-stimulated c-*fos* expression in 6-OHDA-lesioned rats[207] and chronic infusion of the muscarinic antagonist, trihexyphenidyl, augmented 6-OHDA-induced PPD and SP mRNA ipsilateral and contralateral to the lesion.[208] These data support the hypothesis first articulated by Di Chiara and colleagues that endogenous inhibitory cholinergic tone prevents striatonigral neurons from overexcitation during D_1 receptor stimulation and potentially normalizes gene expression after D_1 stimulation.[20a] M_4 receptors probably mediate this cholinergic inhibition because D_1 receptor-stimulated adenylate cyclase mediates neuropeptide and IEG transcription[210] and striatonigral neurons express M_4 receptors[169,170] which are coupled to inhibition of adenylate cyclase activity.[171,172]

3.5 Effects of Muscarinic Cholinergic and D_2 Dopaminergic Inputs Converging on Striatopallidal Neurons

Acetylcholine and dopamine also exert opposite effects on striatopallidal gene expression. However, unlike the tonically active cholinergic regulation of striatonigral gene expression, cholinergic neurotransmission exerts minimal influence on striatopallidal gene expression under normal conditions. This fact was demonstrated by the inability of scopolamine to modify the basal level of striatal PPE mRNA expression.[204] In contrast, after administration of oxotremorine, Fos immunoreactivity in retrogradely labeled striatopallidal neurons[203] and PPE expression[204] were significantly elevated. Furthermore, scopolamine blocked the increase in PPE mRNA induced by either amphetamine[204] or the full D_1 agonist, SKF-82958.[206] However, oxotremorine had no additional effect on PPE expression when co-administered with amphetamine.[204] This lack of an additive effect may indicate that the two drugs affect PPE gene expression via the same ultimate mechanism, i.e., direct muscarinic receptor stimulation by oxotremorine and indirect stimulation of acetylcholine release by amphetamine. The muscarinic receptor which may be involved in facilitating PPE gene expression is most likely the M_1 subtype which is expressed by most, if not all, striatopallidal neurons and is coupled to activation of phosphoinositide hydrolysis. Consistent with these findings are reports that scopolamine significantly attenuates D_2 antagonist-induced enkephalin immunoreactivity,[15,16] PPE mRNA,[17,18,206,211] and FOS immunoreactivity.[212] Thus, D_2 receptors tonically inhibit, whereas D_1 receptors phasically stimulate, PPE gene expression in a manner similar to, and partially mediated by, D_1/D_2 regulation of acetylcholine release.

Finally, reciprocal regulation of cholinergic neurons by enkephalin is indicated by the prevalent distribution of delta opioid receptors on cholinergic interneurons and the fact that delta opioid agonists inhibit acetylcholine release.[213] Thus, local enkephalin release may serve as a negative feedback signal to normalize gene expression in striatopallidal neurons in response to increased cholinergic activity. This action may underlie the finding that local infusion of a delta opioid receptor agonist reduced neuroleptic-stimulated c-*fos* and *zif*268 mRNA expression in the dorsal striatum.[214]

3.6   Mechanisms Underlying Cholinergic and Dopamine Interactions

Based on the information described above, a well-orchestrated program of events between striatal medium spiny and cholinergic neurons is hypothesized to occur in response to dopamine stimulation[20a] (see Figure 1). This cell-to-cell sequence starts with stimulation of D_1 receptors located on the soma and dendrites of striatonigral neurons as well as on their terminals in the substantia nigra. Activation of somatodendritic receptors results in multiple cellular responses which include release of SP, GABA, and dynorphin and a compensatory increase in PPD/SP expression. In the striatum, SP release activates neurokinin NK_1 receptors on cholinergic neurons. Acetylcholine released via this route would bind to muscarinic, probably M_4 receptors and inhibit PPD/SP gene expression in striatonigral neurons by decreasing adenylate cyclase-dependent transcription. In contrast, acetylcholine would enhance the expression of PPE in striatopallidal neurons, possibly via M_1 receptors coupled to phosphoinositide hydrolysis. At the level of the substantia nigra, D_1 receptor stimulation, which facilitates local GABA release, would result in disinhibition of glutamate release in the thalamostriatal and corticostriatal pathways (see Section 3.3) and lead to stimulation of acetylcholine release. In this manner, the cholinergic neuron would be in a strategic position to serve as a feedforward inhibitor of psychostimulant-induced striatonigral activity and a facilitator of striatopallidal activity, as exemplified by scopolamine's ability to enhance D_1-stimulated PPD and SP expression and attenuate PPE expression.

To date, most data on the cholinergic regulation of striatal gene expression has been obtained after systemic drug administration. Proving the validity of the above sequence of events requires demonstration that the striatum and the substantia nigra are the sites of acetylcholine/dopamine interaction which are relevant to normal and stimulated gene expression. In addition, the precise contribution of muscarinic receptor subtypes in these events remains to be determined.

The inhibitory pathway from cholinergic interneurons to striatonigral neurons also provides an intrastriatal mechanism for the well-known phenomenon of D_1/D_2 synergy. D_1/D_2 synergy is considered to be important in mediating striatonigral gene expression induced by indirect dopamine agonists. Although D_1 receptors are thought to control striatonigral gene expression, a number of studies demonstrate that D_2 receptor antagonists block stimulant-induced dynorphin immunoreactivity and PPD gene expression[6,7,9-11] or FOS-like immunoreactivity in striatonigral neurons identified by retrograde tracers.[215] Regardless of the controversy over D_1/D_2 receptor segregation on striatonigral and striatopallidal neurons, it seems unlikely that the synergy could occur within the same neuron, because the two receptors are coupled to different G proteins and have opposite effects on the cAMP signaling pathway which is essential for PPD transcription.[216] Thus, the contribution of D_2 receptor stimulation to the full expression of D_1-initiated action *in vivo* is likely to include an indirect, transynaptic mechanism. Because D_2 tone is an important inhibitory force on acetylcholine release, it is possible that concomitant stimulation of D_2 receptors, by minimizing acetylcholine release, decreases cholinergic inhibition of striatonigral neurons and, thus, synergistically enhances D_1 stimulation of these neurons. In contrast, D_2 receptor blockade, by increasing acetylcholine release, would attenuate D_1-stimulated gene expression. In support of the acetylcholine-dependent D_2 contribution, we recently found that the D_2-selective antagonist, eticlopride, blocked D_1 agonist-stimulated PPD and SP mRNA expression in the rat striatum and that the effect of eticlopride was completely prevented by scopolamine.[206]

Finally, the influence of muscarinic receptor tone on presynaptic neurotransmitter release must be considered. Systemic scopolamine administration increases the extracellular concentration of striatal dopamine,[217,218] indicating an inhibitory influence of muscarinic neurotransmission on striatal dopamine release. This cholinergic inhibition is probably mediated by M_2 receptors which reside on dopamine terminals.

There also is evidence that muscarinic receptor stimulation inhibits the excitatory response of striatal neurons to cortical stimulation *in vivo*[219] or intrastriatal stimulation *in vitro*.[220] Thus, it would be reasonable to speculate that there are presynaptic receptors on corticostriatal terminals that mediate the inhibitory effects of acetylcholine on glutamatergic inputs to the striatum. This presynaptic action may be an additional way by which acetylcholine inhibits striatonigral gene expression.

3.7 Summary

Striatal cholinergic transmission is organized anatomically to serve as an important intrinsic modulator which balances the influence of dopaminergic inputs on striatonigral and striatopallidal neuronal activities, including gene expression. Under normal conditions, cholinergic inhibition and dopaminergic stimulation interact to determine the basal activity of striatonigral gene expression. Conversely, dopaminergic inhibition interacts with cholinergic stimulation to determine basal striatopallidal gene expression. Furthermore, the activity level of cholinergic neurons is subject to reciprocal control by both dopaminergic and medium spiny neuronal inputs, in addition to glutamatergic inputs. During dopamine receptor stimulation, cholinergic activity is balanced by D_1 receptor stimulation in the striatum and substantia nigra and dampened by D_2 receptor-triggered inhibition. The dose of drug and the acute vs. chronic nature of psychostimulant administration will determine whether the balance is tipped toward D_1 or D_2 tone. Apparently, cholinergic interneurons play a buffer-like role in normalizing responses of their neighboring projection neurons to changes in dopaminergic activity.

4. CONCLUSION: GLUTAMATE AND ACETYLCHOLINE REGULATE THE RESPONSES OF MEDIUM SPINY NEURONS TO DOPAMINE

The following is a model of how dopaminergic, glutamatergic, and muscarinic neurotransmission may interact to regulate medium spiny neuronal gene expression induced by psychostimulants[20a] (see Figure 1). First, dopamine receptor agonists increase glutamate release in the dorsal striatum primarily via the polysynaptic striatonigral-thalamostriatal and thalamocortical-striatal circuitry. Glutamate then would stimulate the release of acetylcholine which would dampen the response of striatonigral neurons to dopaminergic stimulation and enhance that of the striatopallidal neurons. Because cholinergic interneurons fire spontaneously and their membrane potentials fall within 2 mV of action potential threshold,[167] they are much more sensitive to subtle changes in glutamate receptor tone than are medium spiny neurons. Therefore, changes in glutamate receptor activity may preferentially impact cholinergic neurons and possibly other striatal interneurons, which translate those changes to medium spiny neurons. In the presence of EAA receptor antagonists, dopaminergic neuronal activity may be attenuated by disinhibition of striatonigral GABA release (see Section 2.4). This disinhibition could occur by turning off an inhibitory interneuron in the striatum, possibly the GABA/parvalbumen-containing neuron, which exerts powerful control over striatonigral activity. Under these conditions, when a psychostimulant is administered, the nigrostriatal neurons would be incapable of responding with adequate dopamine release and behavioral effects would be attenuated. A corollary of this decreased dopamine release would be the blockade by EAA antagonists of IEG and neuropeptide mRNA induction in striatonigral neurons. Furthermore, repeated administration of EAA antagonists would be predicted to prevent the initiation of long-term genomic alterations which underlie the changes in neuronal responsiveness required for the induction of behavioral sensitization. This line of reasoning promotes the idea that EAA receptor antagonists may prove to be potentially useful in the search for novel therapeutic approaches to drug abuse.

Blockade of muscarinic receptors with acute scopolamine augments, whereas repeated scopolamine administration prevents, psychostimulant-induced behavioral sensitization (see Section 3.2). However, in contrast to EAA antagonists, acute and chronic administration of muscarinic antagonists augments striatonigral gene expression (see Section 3.4). The acute effect of and the effect of repeated low doses of scopolamine on behavioral sensitization may be attributed to presynaptic augmentation of dopamine release (see Section 3.6). However, blockade of behavioral sensitization by high doses of scopolamine may be explained most easily by proposing a novel mechanism for a postsynaptic muscarinic antagonist effect that depends on augmented kappa opioid receptor stimulation. Psychostimulant-induced neuropeptide and IEG induction is interpreted to mean that the striatonigral pathway is activated. This concept is supported by evidence that a D_1 agonist increases extracellular levels of dynorphin in the striatum and substantia nigra,[221] which should increase kappa receptor stimulation. Kappa agonists have been demonstrated to attenuate basal[222] and cocaine-stimulated[223] dopamine release in the dorsal and ventral striatum and to attenuate the acute and behavioral sensitizing effects of cocaine.[224] Therefore, the effects of repeated high doses of scopolamine, in contrast to those of EAA antagonists, may require sustained augmentation of the endogenous kappa opioid system in order for postsynaptic muscarinic receptor blockade to affect psychostimulant-induced behavioral sensitization. Future studies will determine whether or not this model of EAA and muscarinic interactions with dopamine is tenable.

ACKNOWLEDGMENT

The studies described in this review by the authors were supported by DA03982.

REFERENCES

1. Gerfen, C.R., Engber, T.M., Mahan, L.C., Susel, Z., Chase, T.N., Monsma, F.J., Jr. and Sibley, D.R., D_1 and D_2 dopamine receptor-regulated gene expression of striatonigral and striatopallidal neurons, *Science*, 250, 1429, 1990.

2. Le Moine, C., Normand, E., Guitteny, A.F., Fouque, B., Teoule, R. and Bloch, B., Dopamine receptor gene expression by enkephalin neurons in rat forebrain, *Proc. Natl. Acad. Sci. U.S.A.*, 87, 230, 1990.

3. Le Moine, C. and Bloch, B., D1 and D2 dopamine receptor gene expression in the rat striatum: sensitive cRNA probes demonstrate prominent segregation of D1 and D2 mRNAs in distinct neuronal populations of the dorsal and ventral striatum, *J. Comp. Neurol.*, 355, 418, 1995.

4. Cole, A. J., Bhat, R. V., Patt, C., Worley, P. F. and Baraban, J. M., D_1 dopamine receptor activation of multiple transcription factor genes in rat striatum, *J. Neurochem.*, 58, 1420, 1992.

5. Moratalla, R., Robertson, H. A. and Graybiel, A. M., Dynamic regulation of NGFI-A (*zif*268, egr1) gene expression in the striatum, *J. Neurosci.*, 12, 2609, 1992.

6. Sivam, S.P., Cocaine selectively increases striatonigral dynorphin levels by a dopaminergic mechanism, *J. Pharmacol. Exp. Ther.*, 250, 818, 1989.

7. Smiley, P.L., Johnson, M., Bush, L., Gibb, J.W. and Hanson, G.R., Effects of cocaine on extrapyramidal and limbic dynorphin systems, *J. Pharmacol. Exp. Ther.*, 253, 938, 1990.

8. Surmeier, D.J., Reiner, A., Levine, M.S. and Ariano, M.A., Are neostriatal dopamine receptors co-localized?, *TINS*, 16, 299, 1993.

9. Wang, J. Q. and McGinty, J. F., Differential effects of D_1 and D_2 dopamine receptor antagonists on acute amphetamine- or methamphetamine-induced upregulation of *zif*268 mRNA expression in rat striatum, *J. Neurochem.*, 65, 2706, 1995.

10. Wang, J. Q. and McGinty, J. F., D_1 and D_2 receptor regulation of preproenkephalin and preprodynorphin mRNA in rat striatum following acute injection of amphetamine and methamphetamine, *Synapse*, 22, 114, 1996.

11. Daunais, J. B. and McGinty, J. F., The effects of D_1 or D_2 dopamine receptor blockade on *zif*268 and preprodynorphin gene expression in rat forebrain following a short-term cocaine binge, *Mol. Brain Res.*, 35, 237, 1996.

12. Bannon, M. J., Kelland, M. and Chiodo, L., Medial forebrain bundle stimulation or D-2 dopamine receptor activation increases preproenkephalin mRNA in rat striatum, *J. Neurochem.*, 52, 859, 1989.

13. Hurd, Y. L. and Herkenham, M., Influence of a single injection of cocaine, amphetamine or GBR 12909 on mRNA expression of striatal neuropeptides, *Mol. Brain Res.*, 16, 97, 1992.

14. Steiner, H. and Gerfen, C. R., Cocaine-induced c-*fos* messenger RNA is inversely related to dynorphin expression in striatum, *J. Neurosci.*, 13, 5066, 1993.

15. Hong, J. S., Yang, H. T. T., Gillin, J. C. and Costa, E., Effects of long-term administration of antipsychotic drugs on enkephalinergic neurons, *Adv. Biochem. Psychopharmacol.*, 24, 223, 1980.

16. Young, W. S. III, Bonner, T. I. and Brann, M. R., Mesencephalic dopamine neurons regulate the expression of neuropeptide mRNAs in the rat forebrain, *Proc. Natl. Acad. Sci. U.S.A.*, 83, 9827, 1986.

17. Angulo, J.A., Cadet, J., Woolley, C., Suber, F. and McEwen, B., Effect of chronic typical and atypical neuroleptic treatment on proenkephalin mRNA levels in the striatum and nucleus accumbens of the rat, *J. Neurochem.*, 54, 1889, 1990.

18. Augood, S. J., Faull, R. L. and Emson, P. C., Contrasting effects of raclopride and SCH 23390 on the cellular content of proenkephalin A mRNA in rat striatum: a quantitative non-radioactive *in situ* hybridization study, *Eur. J. Pharmacol.*, 4, 102, 1992.

19. Jiang, H. K., McGinty, J. F. and Hong, J. S., Differential modulation of striatonigral dynorphin and enkephalin by dopamine receptor subtypes, *Brain Res.*, 507, 57, 1990.

20. Li, S. J., Jiang, H. K., Stachowiak, M. S., Hudson, P. M., Owyang, V., Nanry, K., Tilson, H. A. and Hong, J. S., Influence of nigrostriatal dopaminergic tone on the biosynthesis of dynorphin and enkephalin in rat striatum, *Mol. Brain Res.*, 8, 219, 1990.

20a. Di Chiara, G., Morelli, M. and Consolo, S., Modulatory functions of neurotransmitters in the striatum: Ach/dopamine/NMDA interactions, *Trends Neurosci.*, 17, 228, 1994.

21. Dube, L., Smith, A. D. and Bolam, J. P., Identification of synaptic terminals of thalamic or cortical origin in contact with distinct medium size spiny neurons in the rat neostriatum, *J. Comp. Neurol.*, 267, 455, 1988.

22. Frotscher, M., Rinner, U., Hassler, R. and Wagner, A., Termination of cortical afferents on identified neurons in the caudate nucleus of the cat: a combined Golgi-electron microscope degeneration study, *Exp. Brain Res.*, 41, 329, 1981.

23. Smith, Y., Bennett, B. D., Bolam, J. P., Parent, A. and Sadikot, A. F., Synaptic relationships between dopaminergic afferents and cortical or thalamic input in the sensorimotor territory of the striatum in monkey, *J. Comp. Neurol.*, 344, 1, 1994.

24. Somogyi, P., Bolam, J. P. and Smith, A. D., Monosynaptic cortical input and local axon collaterals of identified striatonigral neurons. A light and electron microscopic study using the Golgi-peroxidase transport-denegeration procedure, *J. Comp. Neurol.*, 195, 567, 1981.

25. Lapper, S. R. and Bolam, J. P., Input from the frontal cortex and the parafascicular nucleus to cholinergic interneurons in the dorsal striatum of the rat, *Neuroscience*, 51, 533, 1992.

26. Meredith, G. E. and Wouterlood, F. G., Hippocampal and midline thalamic fibres and terminals in relation to the choline acetyltransferase-immunoreactive neurons in nucleus accumbens of the rat: a light and electron microscopic study, *J. Comp. Neurol.*, 296, 204, 1990.

27. Freund, T. F., Powell, J. F. and Smith, A. D., Tyrosine hydroxylase-immunoreactive boutons in synaptic contact with identified striatonigral neurons, with particular reference to dendritic spines, *Neuroscience*, 13, 1189, 1984.

28. Pickel, V. M., Chan, J. and Sesack, S. R., Cellular basis for interactions between catecholaminergic afferents and neurons containing Leu-enkephalin-like immunoreactivity in rat caudate-putamen nuclei, *J. Neurosci. Res.*, 31, 212, 1992.

29. Bouyer, J. J., Park, D. H., Joh, T. H. and Pickel, V. M., Chemical and structural analysis of the relation between cortical inputs and tyrosine hydroxylase-containing terminals in rat neostriatum, *Brain Res.*, 302, 267, 1984.

30. Smith, A. D. and Bolam, J. P., The neural network of the basal ganglia as revealed by the study of synaptic connections of identified neurons, *TINS*, 13, 259, 1990.

31. Sesack, S. R. and Pickel, V. M., In the rat medial nucleus accumbens, hippocampal and catecholaminergic terminals converge on spiny neurons and are in apposition to each other, *Brain Res.*, 527, 266, 1990.

32. Totterdell, S. and Smith, A. D., Convergence of hippocampal and dopaminergic input onto identified neurons in the nucleus accumbens of the rat, *J. Chem. Neuroanat.*, 2, 285, 1989.

33. Monaghan, D., Bridages, R. and Cotman, C., The excitatory amino acid receptors: their classes, pharmacology, and distinct properties in the function of the central nervous system, *Annu. Rev. Pharmacol. Toxicol.*, 29, 365, 1989.

34. Albin, R. L., Makowiec, R. L., Hollingsworth, Z. R., Dure IV, L. S., Penney, J. B. and Young, A. B., Excitatory amino acid binding sites in the basal ganglia of the rat: a quantitative autoradiographic study, *Neuroscience*, 46, 35, 1992.

35. Greenamyre, J. T., Olsen, J. M. M., Penney, J. B. and Young, A. B., Autoradiographic characterization of N-methyl-D-aspartate-, quisqualate- and kainate-sensitive glutamate binding sites, *J. Pharmacol. Exp. Ther.*, 233, 254, 1985.

36. Halpain, S., Wieczorek, C. M. and Rainbow, T. C., Localization of L-glutamate receptors in rat brain by quantitative autoradiography, *J. Neurosci.*, 4, 2247, 1984.

37. Monaghan, D. T., Yao, D. and Cotman, C. W., L-[³H]glutamate binds to kainate-, NMDA-, and AMPA-sensitive binding sites: an autoradiographic analysis, *Brain Res.*, 340, 378, 1985.

38. Greenamyre, J. T. and Young, A. B., Synaptic localization of striatal NMDA, quisqualate and kainate receptors, *Neurosci. Lett.*, 101, 133, 1989.

39. Errami, M. and Nieoullon, A., α-[³H]Amino-3-hydroxy-5-methyl-4-isoxazolepropionic acid binding to rat striatal membranes: effects of selective brain lesions, *J. Neurochem.*, 51, 579, 1988.

40. Wullner, U., Testa, C. M., Catania, M. V., Young, A. B. and Penney, J. B., Jr., Glutamate receptors in striatum and substantia nigra: effects of medial forebrain bundle lesions, *Brain Res.*, 645, 98, 1994.

41. Errami, M. and Nieoullon, A., Development of a micromethod to study the Na⁺-independent L-[³H]glutamic acid binding to rat striatal membranes. II. Effects of selective striatal lesions and deafferentations, *Brain Res.*, 366, 178, 1986.

42. Roberts, P. J., McBean, G. J., Sharif, N. A., and Thomas, E. M., Striatal glutamatergic function; modifications following specific lesions, *Brain Res.*, 235, 83, 1982.

43. Tallaksen-Greene, S. J., Wiley, R. G. and Albin, R. L., Localization of striatal excitatory amino acid binding site subtypes to striatonigral projection neurons, *Brain Res.*, 594, 165, 1992.

44. Norman, A. B., Ford, L. M., Kolmonpunporn, M. and Sanberg, P. R., Chronic treatment with MK-801 increases the quinolinic acid-induced loss of D-1 dopamine receptors in rat striatum, *Eur. J. Pharmacol.*, 176, 363, 1990.

45. Tallaksen-Greene, S. J. and Albin, R. L., Localization of AMPA-selective excitatory amino acid receptor subunits in identified population of striatal neurons, *Neuroscience*, 61, 509, 1994.

46. Nakanishi, S., Masu, M., Bessho, Y., Nakajima, Y., Hayashi, Y. and Shigemoto, R., Molecular diversity of glutamate receptors and their physiological functions, *EXS*, 71, 71, 1994.

47. Schoepfer, R., Monyer, H., Sommer, B., Wisden, W., Sprengel, R., Kuner, T., Lomeli, H., Herb, A., Kohler, M. and Burnashev, N., Molecular biology of glutamate receptors, *Prog. Neurobiol.*, 42, 353, 1994.

48. Westbrook, G. L., Glutamate receptor subtypes, *Curr. Opinion Neurobiol.*, 4, 337, 1994.

49. Wisden, W. and Seeburg, P. H., Mammalian ionotropic glutamate receptors, *Curr. Opinion Neurobiol.*, 3, 291, 1993.

50. Gall, C., Sumikawa, K. and Lynch, G., Levels of mRNA for a putative kainate receptor are affected by seizures, *Proc. Natl. Acad. Sci. U.S.A.*, 87, 7643, 1990.

51. Pellegrini-Giampietro, D. E., Bennett, M. V. L. and Zukin, R. S., Differential expression of three glutamate receptor genes in developing rat brain: an *in situ* hybridization study, *Proc. Natl. Acad. Sci. U.S.A.*, 88, 4157, 1991.

52. Sato, K., Kiyama, H. and Tohyama, M., The differential expression patterns of messenger RNAs encoding non-N-methyl-D-aspartate glutamate receptor subunits (GluR1-4) in the rat brain, *Neuroscience*, 52, 515, 1993.

53. Standaert, D. G., Testa, C. M., Young, A. B. and Penney, J. B., Jr., Organization of N-methyl-D-aspartate glutamate receptor gene expression in the basal ganglia of the rat, *J. Comp. Neurol.*, 343, 1, 1994.

54. Landwehrmeyer, G. B., Standaert, D. G., Testa, C. M., Penney, J. B. and Young, A.B., NMDA receptor subunit mRNA expression by projection neurons and interneurons in rat striatum, *J. Neurosci.*, 15, 5297, 1995.

55. Martin, L. J., Blackstone, C. D., Levey, A. I., Huganir, R. L. and Price, D. L., AMPA glutamate receptor subunits are differentially distributed in rat brain, *Neuroscience*, 53, 327, 1993.
56. Petralia, R. S. and Wenthold, R. J., Light and electron immunocytochemical localization of AMPA-selective glutamate receptors in the rat brain, *J. Comp. Neurol.*, 318, 329, 1992.
57. Kendall, D. A., Direct and indirect responses to metabotropic glutamate receptor activation in the brain, *Biochem. Soc. Transact.*, 21, 1120, 1993.
58. Schoepp, D. D. and Conn, P. J., Metabotropic glutamate receptors in brain function and pathology, *TIPS*, 14, 13, 1993.
59. Ohishi, H., Shigemoto, R., Nakanishi, S. and Mizuno, N., Distribution of the messenger RNA for a metabotropic glutamate receptor, mGluR2, in the central nervous system of the rat, *Neuroscience*, 53, 1009, 1993.
60. Ohishi, H., Akazawa, C., Shigemoto, R., Nakankshi, S. and Mizuno, N., Distribution of the mRNAs for L-2-amino-4-phosphonobutyrate-sensitive metabotropic glutamate receptors, mGluR4 and mGluR7, in the rat brain, *J. Comp. Neurol.*, 360, 555, 1995.
61. Shigemoto, R., Nakanishi, S. and Mizuno, N., Distribution of the mRNA for a metabotropic glutamate receptor (mGluR1) in the central nervous system: an *in situ* hybridization study in adult and developing rat, *J. Comp. Neurol.*, 322, 121, 1992.
62. Tanabe, Y., Nomura, A., Masu, M., Shigemoto, R., Mizuno, N. and Nakanishi, S., Signal transduction, pharmacologic properties, and expression patterns of two rat metabotropic glutamate receptors, mGluR3 and mGluR4, *J. Neurosci.*, 13, 1372, 1993.
63. Testa, C. M., Standaert, D. G., Young, A. B. and Penney, J. B., Jr., Metabotropic glutamate receptor mRNA expression in the basal ganglia of the rat, *J. Neurosci.*, 14, 3005, 1994.
64. Kinzie, J. M., Saugstad, J. A., Westbrook, G. L. and Segerson, T. P., Distribution of metabotropic glutamate receptor 7 messenger RNA in the developing and adult rat brain, *Neuroscience*, 69, 167, 1995
65. Martin, L. J., Blackstone, C. D., Huganir, R. L. and Price, D. L., Cellular localization of a metabotropic glutamate receptor in rat brain, *Neuron*, 9, 259, 1992.
66. Shigemoto, R., Nomura, S., Ohishi, H., Sugihara, H., Nakanishi, S. and Mizuno, N., Immunohistochemical localization of a metabotropic glutamate receptor, mGluR5, in the rat brain, *Neurosci. Lett.*, 163, 53, 1993.
67. Fotuhi, M., Sharp, A. H., Glatt, C. E., Hwang, P. M., von Krosigk, M., Snyder, S. H. and Dawson, T. M., Differential localization of phosphoinositide-linked metabotropic glutamate receptor (mGluR1) and the inositol 1,4,5,-trisphosphate receptor in rat brain, *J. Neurosci.*, 13, 2001, 1993.
68. Romano, C., Sesma, M. A., McDonald, C. T., O'Malley, K., Van Der Pol, A. N. and Olney, J. W., Distribution of metabotropic glutamate receptor mGluR5 immunoreactivity in rat brain, *J. Comp. Neurol.*, 355, 455, 1995.
69. Karler, R., Calder, L. D., Chaudhry, I. A. and Turkanis, S. A., Blockade of "reverse tolerance" to cocaine and amphetamine by MK-801, *Life Sci.*, 45, 599, 1989.
70. Karler, R., Chaudhry, I. A., Calder, L. D. and Turkanis, S. A., Amphetamine sensitization and the excitatory amino acids, *Brian Res.*, 537, 76, 1990.
71. Stewart, J. and Druhan, J. P., The development of both conditioning and sensitization of the behavioral activating effects of amphetamine is blocked by the noncompetitive NMDA receptor antagonist, MK-801, *Psychopharmacology*, 110, 125, 1993.
72. Wolf, M. E. and Khansa, M. R., Repeated administration of MK-801 produces sensitization to its own locomotor stimulation effects but blocks sensitization to amphetamine, *Brain Res.*, 562, 164, 1991.

73. Wolf, M. E. and Jeziorski, M., Coadministration of MK-801 with amphetamine, cocaine or morphine prevents rather than transiently masks the development of behavioral sensitization, *Brain Res.*, 613, 291, 1993.

74. Ohmori, T., Abekawa, A., Muraki, A. and Koyama, T., Competitive and noncompetitive NMDA antagonists block sensitization to methamphetamine, *Pharmacol. Biochem. Behav.*, 48, 587, 1994.

75. Karler, R., Calder, L. D. and Bedingfield, J. B., Cocaine behavioral sensitization and the excitatory amino acids, *Psychopharmacology*, 115, 305, 1994.

76. Karler, R., Calder, L. D. and Turkanis, S. A., DNQX blockade of amphetamine behavioral sensitization, *Brain Res.*, 552, 295, 1991.

77. Bristow, L. J., Thorn, L., Tricklebank, M. D. and Hutson, P. H., Competitive NMDA receptor antagonists attenuate the behavioral and neurochemical effects of amphetamine in mice, *Eur. J. Pharmacol.*, 264, 353, 1994.

78. Karler, R., Calder, L. D., Thai, L. H. and Bedingfield, J. B., A dopaminergic-glutamatergic basis for the action of amphetamine and cocaine, *Brain Res.*, 658, 8, 1994.

79. Witkin, J. M., Blockade of the locomotor stimulant effects of cocaine and methamphetamine by glutamate antagonists, *Life Sci.*, 53, 405, 1993.

80. Hamilton, M. H., de Belleroche, J. S., Gardiner, I. M. and Herberg, L. J., Stimulatory effect of N-methyl-d-aspartate on locomotor activity and transmitter release from rat nucleus accumbens, *Pharmacol. Biochem. Behav.*, 25, 943, 1986.

81. Karler, R., Calder, L. D., Thai, L. H. and Bedingfield, J. B., The dopaminergic, glutamatergic, GABAergic bases for the action of amphetamine and cocaine, *Brain Res.*, 671, 100, 1995.

82. Kelley, A. E. and Throne, L. C., NMDA receptors mediate the behavioral effects of amphetamine infused into the nucleus accumbens, *Brain Res. Bull.*, 29, 247, 1992.

83. Pulvirenti, L., Swerdlow, N. R. and Koob, G. F., Microinjection of glutamate antagonist into the nucleus accumbens reduces psychostimulant locomotion in rats, *Neurosci. Lett.*, 103, 213, 1989.

84. Willins, D. L., Wallace, L. J., Miller, D. D. and Uretsky, N. J., α-Amino-3-hydroxy-5-methylisoxazole-4-propionate/kainate receptor antagonists in the nucleus accumbens and ventral pallidum decrease the hypermotility response to psychostimulant drugs, *J. Pharmacol. Exp. Ther.*, 260, 1145, 1992.

85. Arnt, J., Hyperactivity following injection of a glutamate agonist and 6,7-ADTN into rat nucleus accumbens and its inhibition by THIP, *Life Sci.*, 28, 1597, 1981.

86. Donzanti, B. A. and Uretsky, N. J., Antagonism of the hypermotility response induced by excitatory amino acids in the rat nucleus accumbens, *Naunyn Schmiedeberge's Arch. Pharmacol.*, 325, 1, 1984.

87. Hooks, M. S. and Kalivas, P. W., Involvement of dopamine and excitatory amino acid transmission in novelty-induced motor activity, *J. Pharmacol. Exp. Ther.*, 269, 976, 1994.

88. Thanos, P. K., Jhamandas, K. and Beninger, R. J., N-methyl-D-aspartate unilaterally injected into the dorsal striatum of rats produces contralateral circling: antagonism by 2-amino-7-phosphonoheptamoic acid and cis-flupenthixol, *Brain Res.*, 589, 55, 1992.

89. Sacaan, A. I., Monn, J. A. and Schoepp, D. D., Intrastriatal injection of a selective metabotropic excitatory amino acid receptor agonist induces contralateral turning in the rat, *J. Pharmacol. Exp. Ther.*, 259, 1366, 1991.

90. Sacaan, A. I., Bymaster, F. P. and Schoepp, D. D., Metabotropic glutamate receptor activation produces extrapyramidal motor system activation that is mediated by striatal dopamine, *J. Neurochem.*, 59, 245, 1992.

91. McDonald, J. W., Fix, A. S., Tizzano, J. P. and Schoepp, D. D., Seizures and brain injury in neonatal rats induced by 1S,3R-ACPD, a metabotropic glutamate receptor agonist, *J. Neurosci.*, 13, 4445, 1993.

92. Nestler, E. J., Molecular neurobiology of drug addiction, *Neuropsychopharmacology*, 11, 77, 1994.

93. Johnson, M., Bush, L. G., Gibb, J. W. and Hanson, G. R. Blockade of the 3,4-methylenedioxymethamphetamine-induced changes in neurotensin and dynorphin A system, *Eur. J. Pharmacol.*, 193, 367, 1991.

94. Johnson, M., Bush, L. G., Gibb, J. W. and Hanson, G. R., Role of N-methyl-D-aspartate (NMDA) receptors in the response of extrapyramidal neurotension and dynorphin A system to cocaine and GBR 12909, *Biochem. Pharmacol.*, 41, 649, 1991.

95. Singh, N. A., Midgley, L. P., Bush, L. G., Gibb, J. W. and Hanson, G. R., N-Methyl-D-aspartate receptors mediate dopamine-induced changes in extrapyramidal and limbic dynorphin systems, *Brain Res.*, 555, 233, 1991.

96. Dragunow, M., Logan, B. and Laverty, R., 3,4-methylenedioxymethamphetamine induces Fos-like proteins in rat basal ganglia: reversal with MK-801, *Eur. J. Pharmacol.*, 206, 255, 1991.

97. Ohno, M., Yoshida, H. and Watanabe, S., NMDA receptor-mediated expression of Fos protein in the rat striatum following methamphetamine administration: relation to behavioral sensitization, *Brain Res.*, 665, 135, 1994.

98. Snyder-Keller, A. M., Striatal c-*fos* induction by drugs and stress in neonatally dopamine-depleted rats given nigral transplants: importance of NMDA activation and relevance to sensitization phenomena, *Exp. Neurol.*, 113, 155, 1991.

99. Torres, G. and Rivier, C., Cocaine-induced expression of striatal c-*fos* in the rat is inhibited by NMDA receptor antagonists, *Brain Res. Bull.*, 30, 173, 1993.

100. Cenci, M. A. and Bjorklund, A., Transection of corticostriatal afferents reduces amphetamine- and apomorphine-induced striatal Fos expression and turning behavioral in unilaterally 6-hydroxydopamine-lesioned rats, *Eur. J. Neurosci.*, 5, 1062, 1993.

101. Fu, L. and Beckstead, R. M., Cortical stimulation induces Fos expression in striatal neurons, *Neuroscience*, 46, 329, 1992.

102. Smith, A. J. W. and McGinty, J. F., Acute amphetamine or methamphetamine alters opioid peptide mRNA expression in rat striatum, *Mol. Brain Res.*, 21, 359, 1994.

103. Wang, J. Q. and McGinty, J. F., Alterations in striatal *zif* 268, preprodynorphin and preproenkephalin mRNA expression induced by repeated amphetamine administration in rats, *Brain Res.*, 673, 262, 1995.

104. Wang, J. Q. and McGinty, J. F., Dose-dependent alteration in *zif* 268 and preprodynorphin mRNA expression induced by amphetamine and methamphetamine in rat forebrain, *J. Pharmacol. Exp. Ther.*, 273, 909, 1995.

105. Wang, J. Q., Smith, A. J. W. and McGinty, J. F., A single injection of amphetamine or methamphetamine induces dynamic alterations in c-*fos*, *zif* 268 and preprodynorphin mRNA expression in rat forebrain, *Neuroscience*, 68, 83, 1995.

106. Wang, J. Q., Daunais, J. B. and McGinty, J. F., NMDA receptors mediate amphetamine-induced upregulation of *zif* 268 and preprodynorphin mRNA expression in rat striatum, *Synapse*, 18, 343, 1994.

107. Wang, J. Q., Daunais, J. B. and McGinty, J. F., Role of kainate/AMPA receptors in induction of striatal *zif* 268 and preprodynorphin mRNA by a single injection of amphetamine, *Mol. Brain Res.* 27, 118, 1994.

108. Wang, J. Q. and McGinty, J. F., Acute methamphetamine-induced *zif* 268, preprodynorphin and preproenkephalin mRNA expression in rat striatum depends upon activation of NMDA and kainate/AMPA receptors, *Brain Res. Bull.*, 39, 349, 1996.

109. Berretta, S., Robertson, H. A. and Graybiel, A. M., Dopamine and glutamate agonists stimulate neuron-specific expression of Fos-like protein in the striatum, *J. Neurophysiol.*, 68, 767, 1992.

110. Page, K. J. and Everitt, B. J., Transynaptic induction of c-*fos* in basal forebrain, diencephalic and midbrain neurons following AMPA-induced activation of the dorsal and ventral striatum, *Exp. Brain Res.*, 93, 399, 1993.

111. Vaccarino, F. M., Hayward, M. D., Nestler, E. J., Duman, R. S. and Tallman, J. F., Differential induction of immediate early genes by excitatory amino acid receptor types in primary cultures of cortical and striatal neurons, *Mol. Brain Res.*, 12, 233, 1992.

112. Beckstead, R. M., N-Methyl-D-aspartate acutely increases proenkephalin mRNA in the rat striatum, *Synapse*, 21, 342, 1995.

113. Somers, D. L. and Beckstead, R. M., N-methyl-D-aspartate receptor antagonism alters substance P and met[5]-enkephalin biosynthesis in neurons of the rat striatum, *J. Pharmacol. Exp. Ther.*, 262, 823, 1992.

114. Somers, D. L. and Beckstead, R. M., Striatal preprotachykinin and preproenkephalin mRNA levels and the levels of nigral substance P and pallidal met[5]-enkephalin depend on corticostriatal axons that use the excitatory amino acid neurotransmitters aspartate and glutamate: quantitative radioimmunocytochemical and *in situ* hybridization evidence, *Mol. Brain Res.*, 8, 143, 1990.

115. Uhl, G. R., Navia, B. and Douglass, J., Differential expression of preproenkephalin and preprodynorphin mRNAs in striatal neurons: high levels of preproenkephalin expression depend on cerebral cortical afferents, *J. Neurosci.* 8, 4755, 1988.

116. Wang, J. Q. and McGinty, J. F., Striatal metabotropic glutamate receptors regulate acute amphetamine-stimulated neuropeptide gene expression in the rat striatum, *Soc. Neurosci.* Abstr. in press.

117. Augood, S. J., Westmore, K., Faull, R. L. M. and Emson, P. C., Neuroleptics and striatal neuropeptide gene expression, *Prog. Brain Res.*, 99, 181, 1993.

118. Ziolkowska, B. and Hollt, V., The NMDA receptor antagonist MK-801 markedly reduces the induction of c-*fos* gene by haloperidol in the mouse striatum, *Neurosci. Lett.*, 156, 39, 1993.

119. Cooper, A. J., Wooller, S. and Mitchell, I. J., Elevated striatal Fos immunoreactivity following 6-hydroxydopamine lesioning of the rat is mediated by excitatory amino acid transmission, *Neurosci. Lett.*, 194, 73, 1995.

120. Hagihara, K., Tsumoto, T., Sato, H. and Hata, Y., Actions of excitatory amino acid antagonists on geniculo-cortical transmission in the cat's visual cortex, *Exp. Brain Res.*, 69, 407, 1988.

121. Miller, K. D., Chapman, B. and Stryker, M. P., Visual responses in adult cat visual cortex depend on N-methyl-D-aspartate receptor, *Proc. Natl. Acad. Sci. U.S.A.*, 86, 5183, 1989.

122. Garcia-Munoz, M., Young, S. J. and Groves, P. M., Terminal excitability of the corticostriatal pathway. II. Regulation by glutamate receptor stimulation, *Brain Res.*, 551, 207, 1991.

123. Worley, P. F., Christy, B. A., Nakabeppu, Y., Bhat, R. V., Cole, A. J. and Baraban, J. M., Constitutive expression of *zif* 268 in neocortex is regulated by synaptic activity, *Proc. Natl. Acad. Sci. U.S.A.*, 88, 5106, 1991.

124. Freed, W. J., Glutamatergic mechanisms mediating stimulant and antipsychotic drug effects, *Neurosci. Biobehav. Rev.*, 18, 111, 1994.

125. Yamamoto, B. K. and Davy, S., Dopaminergic modulation of glutamate release in striatum as measured by microdialysis, *J. Neurochem.*, 58, 1736, 1992.

126. Clow, D. W. and Jhamandas, K., Characterization of L-glutamate action on the release of endogenous dopamine from the rat caudate-putamen, *J. Pharmacol. Exp. Ther.*, 248, 722, 1989.

127. Kalivas, P. W. and Duffy, P., D$_1$ receptors modulate glutamate transmission in the ventral tegmental area, *J. Neurosci.*, 15, 5379, 1995.

128. Reid, M. S., Herrera-Marschitz, M., Kehr, J. and Ungerstedt, U., Striatal dopamine and glutamate release: effects of intranigral injections of substance P, *Acta Physiol. Scand.*, 140, 527, 1990.

129. Exposito, I., Sanz, B., Porras, A. and Mora, F., Effects of apomorphine and L-methionine sulphoximine on the release of excitatory amino acid neurotransmitters and glutamine in the striatum of the conscious rat, *Eur. J. Neurosci.*, 6, 287, 1994.

130. Mora, F. and Porras, A., Effects of amphetamine on the release of excitatory amino acid neurotransmitters in the basal ganglia of the conscious rat, *Can. J. Physiol. Pharmacol.*, 71, 348, 1993.

131. Smith, J. A., Mo, Q., Guo, H., Kunko, P. M. and Robinson, S. E, Cocaine increases extraneuronal levels of aspartate and glutamate in the nucleus accumbens, *Brain Res.*, 683, 264, 1995.

132. Nash, J. F. and Yamamoto, B. K., Methamphetamine neurotoxicity and striatal glutamate release: comparison to 3,4-methlenedioxymethamphetamine, *Brain Res.*, 581, 237, 1992.

133. Nash, J. F. and Yamamoto, B. K., Effect of D-amphetamine on the extracellular concentrations of glutamate and dopamine in iprindole-treated rats, *Brain Res.*, 627, 1, 1993.

134. Wang, Z. and Rebec, G. V., Neuronal and behavioral correlates of intrastriatal infusions of amphetamine in freely moving rats, *Brain Res.*, 627, 79, 1993.

135. Keefe, K. A., Zigmond, M. J. and Abercrombie, E. D., Extracellular dopamine in striatum: influence of nerve impulse activity in medial forebrain bundle and local glutamatergic input, *Neuroscience*, 47, 325, 1992.

136. Martinez-Fong, D., Rosales, M. G., Gongora-Alfaro, J. L., Hernandez, S. and Aceves, J., NMDA receptor mediates dopamine release in the striatum of unanesthetized rats as measured by brain microdialysis, *Brain Res.*, 595, 309, 1992.

137. Moghaddam, B., Gruen, R. J., Roth, R. H., Bunney, B. S. and Adams, R. N., Effect of L-glutamate on the release of striatal dopamine: *in vivo* dialysis and electrochemical studies, *Brain Res.*, 518, 55, 1990.

138. Morari, M., O'Connor, W. T., Ungerstedt, U. and Fuxe, K., N-methyl-D-aspartic acid differentially regulates extracellular dopamine, GABA, and glutamate levels in the dorsolateral neostriatum of the halothane-anesthetized rat: an *in vivo* microdialysis study, *J. Neurochem.*, 60, 1884, 1993.

139. Morari, M., O'Connor, W. T., Ungerstedt, U. and Fuxe, K., Dopamine D1 and D2 receptor antagonism differentially modulates stimulation of striatal neurotransmitter levels by N-methyl-D-aspartic acid, *Eur. J. Pharmacol.*, 256, 23, 1994.

140. Moghaddam, B. and Bolinao, M. L., Glutamatergic antagonists attenuate ability of dopamine uptake blockers to increase extracellular levels of dopamine: implication for tonic influence of glutamate on dopamine release, *Synapse*, 18, 337, 1994.

141. Pap, A. and Bradberry, C. W., Excitatory amino acid antagonists attenuate the effects of cocaine on extracellular dopamine in the nucleus accumbens, *J. Pharmacol. Exp. Ther.*, 274, 127, 1995.

142. Weihmuller, F. B., O'Dell, S. J., Cole, B. N. and Marshall, J. F., MK-801 attenuates the dopamine-releasing but not the behavioral effects of methamphetamine: an *in vivo* microdialysis study, *Brain Res.*, 549, 230, 1991.

142a. Keefe, K.A., Zigmond, M.J. and Abercombie, E.D., *In vivo* regulation of extracellular dopamine in the neostriatum: influence of impulse activity and local excitatory amino acids, *J. Neural. Trans.*, 91, 223, 1993.

143. Kawaguchi, Y., Wilson, C. J., Augood, S. J. and Emson, P. C., Striatal interneurons: chemical, physiological and morphological characterization, *TINS*, 18, 527, 1995.

144. Yelnik, J., Percheron, G., Francois, C. and Garnier, A., Cholinergic neurons of the rat and primate striatum are morphologically different, *Prog. Brain Res.*, 99, 25, 1993.

145. Bolam, J. P., Wainer, B. H. and Smith, A. D., Characterization of cholinergic neurons in the rat neostriatum. A combination of choline acetyltransferase immunocytochemistry, Golgi impregnation and electron microscopy, *Neuroscience*, 12, 711, 1984.

146. Phelps, P. E., Houser, C. R. and Vaughn, J. E., Immunocytochemical localization of choline acetyltransferase within the rat neostriatum: a correlated light and electron microscopic study of cholinergic neurons and synapses, *J. Comp. Neurol.*, 238, 286, 1985.

147. Graybiel, A. M., Baugham, R. W. and Eckenstein, F., Cholinergic neuropil of the striatum observes striosomal boundaries, *Nature*, 323, 1986.

148. Lauterborn, J. C., Isackson, P. J., Montalvo, R. and Gall, C. M., *In situ* hybridization localization of choline acetyltransferase mRNA in adult rat brain and spinal cord, *Mol. Brain Res.*, 17, 59, 1993.

149. Meredith, G. E., Blank, B. and Groenwegen, H. J., The distribution and compartmental organization of the cholinergic neurons in nucleus accumbens of the rat, *Neuroscience*, 31, 327, 1989.

150. Izzo, P. N. and Bolam, J. P., Cholinergic synaptic input to different parts of spiny striatonigral neurons in the rat, *J. Comp. Neurol.*, 269, 219, 1988.

151. Chang, H. T., Dopamine-acetylcholine interaction in the rat striatum: a dual-labeling immunocytochemical study, *Brain Res. Bull.*, 21, 295, 1988.

152. Dimova, R., Vullet, J., Nieoullon, A. and Goff, L. K. L., Ultrastructural features of the choline acetyltransferase-containing neurons and relationship with nigral dopaminergic and cortical afferent pathways in the rat striatum, *Neuroscience*, 53, 1059, 1993.

153. Lapper, S. R. and Bolam, J. P., Input from the frontal cortex and the parafascicular nucleus to cholinergic interneurons in the dorsal striatum of rat, *Neuroscience*, 51, 533, 1992.

154. Meredith G. E. and Wouterlood F. G., Hippocampal and midline thalamic fibres and terminals in relation to the choline acetyltransferase-immunoreactive neurons in nucleus accumbens of the rat: a light and electron microscopic study, *J. Comp. Neurol.*, 296, 204, 1990.

155. Aubry, J. M., Schulz, M. F., Pagliusi, S., Schulz, P. and Kiss, J. Z., Coexpression of dopamine D$_2$ and substance P (neurokinin-1) receptor messenger RNAs by a subpopulation of cholinergic neurons in the rat striatum, *Neuroscience*, 53, 417, 1993.

156. Le Moine, C., Tison, F. and Bloch, B., D$_2$ dopamine receptor gene expressed by cholinergic neurons in the rat striatum, *Neurosci. Lett.*, 117, 248, 1990.

157. Guennoun, R. and Bloch, B., Ontogeny of D1 and DARPP-32 gene expression in the rat striatum: an *in situ* hybridization study, *Mol. Brain Res.*, 12, 131, 1992.

158. Parent, A., Extrinsic connections of the basal ganglia, *TINS*, 13, 254, 1990.

159. Elde, R., Schalling, M., Ceddatelli, S., Nakanishi, S. and Hokeflt, T., Localization of neuropeptide receptor mRNA in rat brain: initial observations using probes for neurotensin and substance P receptors, *Neurosci. Lett.*, 120, 134, 1990.

160. Gerfen C. R., Substance P (neurokinin-1) receptor mRNA is selectively expressed in cholinergic neurons in the striatum and basal forebrain, *Brain Res.*, 556, 165, 1991.

161. Maeno, H., Kiyama, H. and Tohyama, M., Distribution of the substance P receptor (NK1-receptor) in the central nervous system, *Mol. Brain Res.*, 18, 43, 1993.

162. Le Moine, C. B., Kieffer, C., Gaveriaux-Ruff, K. B. and Bloch, B., Delta-opioid receptor gene expression in the mouse striatum: Localization in cholinergic neurons of the striatum, *Neuroscience*, 62, 635, 1994.

163. Mansour, A., Thompson, R. C., Akil, H. and Watson, S. J., Delta opioid receptor mRNA distribution in the brain: comparison to delta receptor binding and proenkephalin mRNA, *J. Chem. Neuroanat.*, 6, 351, 1993.

164. Pickel, V. M. and Chan, J., Spiny neurons lacking choline acetyltransferase immunoreactivity are major targets of cholinergic and catecholaminergic terminals in rat striatum, *J. Neurosci. Res.*, 25, 263, 1990.

165. Chang, H. T. and Kita, H., Interneurons in the rat striatum: relationships between parvalbumin neurons and cholinergic neurons, *Brain Res.*, 574, 307, 1992.

166. Vuillet, J., Dimova, R., Nieoullon, A. and Kerkerian-Le Goff, L., Ultrastructural relationships between choline acetyltransferase- and neuropeptide Y-containing neurons in the rat striatum, *Neuroscience*, 46, 351, 1992.

167. Wilson, C. J., Chang, H. T. and Kitai, S. T., Firing patterns and synaptic potentials of identified giant aspiny interneurons in the rat neostriatum, *J. Neurosci.*, 10, 508, 1990.

168. Akaike, A., Sasa, M. and Takaori, S., Muscarinic inhibition as a dominant role in cholinergic regulation of transmission in the caudate nucleus, *J. Pharmacol. Exper. Thera.*, 246, 1129, 1988.

169. Bernard, V., Normand, E. and Bloch, B., Phenotypical characterization of the rat striatal neurons expressing muscarinic receptor gene, *J. Neurosci.*, 12, 3591, 1992.

170. Weiner, D. M., Levey, A. and Brann, M. R., Expression of muscarinic acetylcholine and dopamine receptor mRNAs in rat basal ganglia, *Proc. Natl. Acad. Sci. U.S.A.*, 87, 7050, 1990.

171. Hulme, E. C., Birdsall, N. J. M. and Buckley, N. J., Muscarinic receptor subtypes. *Annu. Rev. Pharmacol. Toxicol.*, 30, 633, 1990.

172. McKinney, M., Muscarinic receptor subtype-specific coupling to second messengers in neuronal systems, *Prog. Brain Res.*, 98, 333, 1993.

173. Hagan, J. J., Tonnaer, J. A. D. M., Rijk, H., Broekkamp, C. L. E. and van Delft, A. M. L., Facilitation of amphetamine-induced rotation by muscarinic antagonists is correlated with M2 receptor affinity, *Brain Res.*, 410, 69, 1987.

174. Shannon, H.E. and Peters, S. C., A comparison of the effects of cholinergic and dopaminergic agents on scopolamine-induced hyperactivity in mice, *J. Pharmacol. Exp. Ther.*, 255, 549, 1990.

175. Arnfred, T. and Randrup, A., Cholinergic mechanism in brain inhibiting amphetamine-induced stereotyped behavior, *Acta Pharmacol. Toxicol.*, 26, 384, 1968.

176. Bymaster, F. P., Heath, I., Hendrix, J. C. and Shannon, H. E., Comparative behavioral and neurochemical activities of cholinergic antagonists in rats, *J. Pharmacol. Exp. Ther.*, 267, 16, 1993.

177. Butkerait, P. and Friedman, E., Scopolamine modulates apomorphine-induced behavior in rats treated with haloperidol or SCH 23390, *Eur. J. Pharmacol.*, 148, 269, 1988.

178. Ohmori, T., Abekawa, T. and Koyama, T., Scopolamine prevents the development of sensitization to methamphetamine, *Life Sci.*, 56, 1223, 1995.

179. Yui, K., Miura, T., Sugiyama, K., Ono, M. and Nagase, M., Methamphetamine plus scopolamine potentiates behavioral sensitization and conditioning, *Eur. J. Pharmacol.*, 279, 135, 1995.

180. Fage, D. and Scatton, B., Opposite effects of D_1 and D_2 receptor antagonists on acetylcholine levels in the rat striatum, *Eur. J. Pharmacol.*, 129, 359, 1986.

181. Ajima, A., Yamaguchi, T. and Kato T., Modulation of acetylcholine release by D_1, D_2 dopamine receptors in rat striatum under freely moving conditions, *Brain Res.*, 518, 193, 1990.

182. Bertorelli, R. and Consolo, S., D_1 and D_2 dopaminergic regulation of acetylcholine release from striata of freely moving rats, *J. Neurochem.*, 54, 2145, 1990.

183. Damsma, G., Tham, C. S., Robertson, G. S. and Fibiger, H. C., Dopamine D_1 receptor stimulation increases striatal acetylcholine release in the rat, *Eur. J. Pharmacol.*, 186, 335, 1990.

184. Damsma, G., Robertson, G. S., Tham, C. S. and Fibiger, H. C., Dopaminergic regulation of striatal cholinergic release: importance of D1 and N-methyl-DM-aspartate receptors, *J. Pharmacol. Exp. Ther.*, 259, 1064, 1991.

185. Consolo, S., Girotti, M., Zambelli, M., Russi, G., Benzi, M. and Bertorelli, R., D_1 and D_2 dopamine receptors and the regulation of striatal acetylcholine release *in vivo*, *Prog. Brain Res.*, 98, 201, 1993.

186. Bertorelli, R., Zambelli, M., Di Chiara, G. and Consolo, S., Reduced endogenous dopamine has different effects on the D_1 and D_2 regulation of *in vivo* ACH release from striatum, *J. Neurochem.*, 59, 353, 1992.

187. Russi, G., Cadoni, C., Di Chiara, G. and Consolo, S., Neuroleptics increase striatal acetylcholine release by a sequential D-1 and D-2 receptor mechanism, *NeuroReport*, 4, 1335, 1993.

188. Florin, S. M., Kuczenski, R. and Segal, D. S., Amphetamine-induced changes in behavior and caudate extracellular acetylcholine, *Brain Res.*, 581, 53, 1992.

189. Lindefors, N., Hurd, Y. L., O'Connor, W. T., Brene, S., Persson, H. and Ungerstedt, U., Amphetamine regulation of acetylcholine and γ-aminobutyric acid in nucleus accumbens, *Neuroscience*, 48, 439, 1992.

190. Acquas, E. and Fibiger, H. C., Concentration dependent effects of neostigmine on d-amphetamine induced increases in striatal acetylcholine release, *Abstr. Soc. Neurosci.*, 21, 2137, 1995.

191. DeBoer, P. and Abercrombie, E. D., Physiological release of striatal acetylcholine *in vivo*: modulation by D1 and D2 dopamine receptor subtypes, *J. Pharmacol. Exp. Ther.*, in press.

192. DeBoer, P. and Abercrombie, E. D., Further characterization of the role of substantia nigra in the modulation of striatal acetylcholine in awake rats, *Abstr. Soc. Neurosci.*, 20, 285, 1994.

193. Furmidge, L. J., Duggan, A. W. and Arbuthnott, G. W., Substance P release from rat nucleus accumbens and striatum: an *in vivo* study using antibody microprobes, *Brain Res.*, 610, 234, 1993.

194. Anderson, J. J., Chase, T. N. and Engber, T. M., Substance P increases release of acetylcholine in the dorsal striatum of freely moving rats, *Brain Res.*, 623, 189, 1993.

195. Anderson, J. J., Randall, S. and Chase, T. N., The neurokinin$_1$ receptor antagonist CP-99,994 reduces catalepsy produced by the dopamine D_2 receptor antagonist raclopride: correlation with extracellular acetylcholine levels in striatum, *J. Pharmacol. Exp. Ther.*, 274, 928, 1995.

196. Arenas, E., Alberch, J., Perez-Navarro, E., Solsona, C. and Marsal, J., Neurokinin receptors differentially mediate endogenous acetylcholine release evoked by tachykinins in the neostriatum, *J. Neurosci.*, 11, 2332, 1991.

197. Petitet, F., Glowinski, J. and Beaujouan, J. C., Evoked release of acetylcholine in the rat striatum by stimulation of tachykinin NK-1 receptors, *Eur. J. Pharmacol.*, 192, 203, 1991.

198. Cameron, D. L. and Williams, J. T., Dopamine D1 receptors facilitate transmitter release, *Nature*, 366, 344, 1993.

199. Baldi, G., Russi, G., Nannini, L., Vezzani, A. and Consolo, S., Trans-synaptic modulation of striatal ACH release *in vivo* by the parafascicular thalamic nucleus, *Eur. J. Neurosci.*, 7, 1117, 1995.

200. Anderson, J. J., Kuo, S. and Chase, T. N., Endogenous excitatory amino acids tonically stimulate striatal acetylcholine release through NMDA but not AMPA receptors, *Neurosci. Lett.*, 176, 264, 1994.

201. Giovannini, M. G., Camilli, F., Mundula, A. and Pepeu, G., Glutamatergic regulation of acetylcholine output in different brain regions: a microdialysis study in the rat, *Neurochem. Int.*, 25, 23, 1994.

202. Zocchi, A. and Pert, A., Alterations in striatal acetylcholine overflow by cocaine, morphine, and MK-801: relationship to locomotor output, *Psychopharmacology*, 115, 297, 1994.

203. Bernard, V., Dumartin, B., Lamy, E. and Bloch, B., Fos immunoreactivity after stimulation or inhibition of muscarinic receptors indicates anatomical specificity for cholinergic control of striatal efferent neurons and cortical neurons in the rat, *Eur. J. Neurosci.*, 5, 1218, 1993.
204. Wang, J. Q. and McGinty, J. F., Muscarinic receptors regulate striatal neuropeptide gene expression in normal and amphetamine-treated rats, *Neuroscience*, in press.
205. Wang, J. Q. and McGinty, J. F., Scopolamine augments c-*fos* and *zif*268 mRNA expression induced by the full D_1 dopamine receptor agonist SKF-82958 in the intact rat striatum, *Neuroscience*, 72, 601, 1996.
206. Wang, J. Q. and McGinty, J. F., Full D1 dopamine receptor agonist SKF 82958 induces neuropeptide mRNA in the normosensitive striatum of rats: regulation of D1/D2 synergy by muscarinic receptors, submitted.
207. Morelli, M., Fenu, S., Cozzolino, A., Pinna, A., Carta, A. and Di Chiara, G., Blockade of muscarinic receptors potentiates D_1 dependent turning behavior and c-*fos* expression in 6-hydroxydopamine-lesioned rats but does not influence D_2 mediated responses, *Neuroscience,* 53, 673, 1993.
208. Mavridis, M., Rogard, M. and Besson, M. J., Chronic blockade of muscarinic cholinergic receptors by systemic trihexyphenidyl (artane[R]) administration modulates but does not mediate the dopaminergic regulation of striatal prepropeptide messenger RNA expression, *Neuroscience*, 66, 37, 1995.
209. Di Chiara, G. and Morelli, M., Dopamine-acetylcholine-glutamate interactions in the striatum, *Adv. Neurol.*, 60, 102, 1993.
210. Hyman, S. E., Cole, R. L., Konradi, C. and Kosofsky, B. E., Dopamine regulation of transcription factor-target interactions in rat striatum, *Chem. Senses,* 20, 257, 1995.
211. Pollack, A. E. and Wooten G. F., D_2 dopaminergic regulation of striatal preproenkephalin mRNA levels is mediated at least in part through cholinergic interneurons, *Mol. Brain Res.*, 13, 35, 1992.
212. Guo, N., Robertson, G. S. and Fibiger, H. C., Scopolamine attenuates haloperidol-induced c-*fos* expression in the striatum, *Brain Res.*, 588, 164, 1992.
213. Mulder, A. H., Wardeh, G., Hogenboom, F. and Frankhuyzen, A. L., κ- and δ-opioid receptor agonists differentially inhibit striatal dopamine and acetylcholine release, *Nature*, 308, 278, 1984.
214. Steiner, H., Mariotti, R. and Gerfen, C. R., Enkephalin inhibits immediate early gene expression induced by blockade of D2 dopamine receptors in striatum, *Abstr. Soc. Neurosci.*, 21, 1904, 1995.
215. Ruskin, D. N. and Marshall, J. F., Amphetamine- and cocaine-induced Fos in the rat striatum depends on D_2 dopamine receptor activation, *Synapse*, 18, 233, 1994.
216. Cole, R. L., Konradi, C., Douglass, J. and Hyman, S. E., Neuronal adaptation to amphetamine and dopamine: molecular mechanisms of prodynorphin gene regulation in rat striatum, *Neuron*, 14, 813, 1995.
217. Blackburn, J. R., Chapman, C. A., Blaha, C. D. and Yeomans, J. S., Striatal dopamine levels increase following peripheral scopolamine injections and following application of scopolamine to the tegmental pedunculopontine nucleus, *Abstr. Neurosci. Soc.*, 20, 1559, 1994.
218. Dewey, S. L., Smith, G. S., Logan, J., Brodie, J. D., Simkowitz, P., MacGregor, R. R., Fowler, J. S., Volkow, N. D. and Wolf, A. P., Effects of central cholinergic blockade on striatal dopamine release measured with positron emission tomography in normal and human subjects, *Proc. Natl. Acad. Sci. U.S.A.*, 90, 11816, 1993.
219. Hsu, K. S., Huang, C. C. and Gean, P. W., Muscarinic depression of excitatory synaptic transmission mediated by the presynaptic M3 receptors in the rat neostriatum, *Neurosci. Lett.*, 197, 141, 1995.

220. Sugita, S., Uchimura, N., Jiang, Z. G. and North, R. A., Distinct muscarinic receptors inhibit release of gamma-aminobutyric acid and excitatory amino acids in mammalian brain, *Proc. Natl. Acad. Sci. U.S.A.*, 88, 2608, 1991.

221. You, Z. B., Herrera-Marschitz, M., Nylander, I., Goiny, M., O'Connor, W. T., Ungerstedt, U. and Terenius, L., The striatonigral dynorphin pathway of the rat studied with *in vivo* microdialysis. II. Effects of dopamine D_1 and D_2 receptor agonists, *Neuroscience*, 63, 427, 1994.

222. Spanagel, R., Herz, A. and Shippenberg, T. S., Opposing tonically active endogenous opioid systems modulate the mesolimbic dopaminergic pathway, *Proc. Natl. Acad. Sci. U.S.A.*, 89, 2046, 1992.

223. Maisonneuve, I. M., Archer, S. and Glick, S. D., U50,488, a kappa opioid receptor agonist, attenuates cocaine-induced increases in extracellular dopamine in the nucleus accumbens of rats, *Neurosci. Lett.*, 181, 57, 1994.

224. Heidbreder, C. A., Goldberg, S. R. and Shipperberg, T. S., The kappa-opioid receptor agonist U-69593 attenuates cocaine-induced behavioral sensitization in the rat, *Brain Res.*, 616, 335, 1993.

Chapter **5**

MOLECULAR MECHANISMS OF STRIATAL GENE REGULATION: A CRITICAL ROLE FOR GLUTAMATE IN DOPAMINE-MEDIATED GENE INDUCTION

Steven E. Hyman, Rebecca L. Cole, Michael Schwarzschild, Douglas Cole, Bruce Hope, and Christine Konradi

CONTENTS

0-8493-8550-4/96/$0.00+$.50
© 1996 by CRC Press, Inc.

1. DRUG-INDUCED ADAPTATIONS IN DOPAMINERGIC CIRCUITS

A central observation concerning many drugs that act on dopamine systems, including the psychostimulant drugs, (e.g., cocaine and amphetamine) the antipsychotic drugs, and the dopamine precursor used in the treatment of Parkinson's disease, L-DOPA, is that their chronic effects on behavior differ markedly from their acute effects. The initial molecular targets for these drugs have been identified and cloned, although some controversy remains with respect to the precise targets of antipsychotic drugs. The primary molecular target of the psychostimulants are the monoamine transporters, of which the dopamine transporter appears to be the most significant for their acute reinforcing effects. The antipsychotic drugs produce their therapeutic effects, as well as many of their side effects, by acting as antagonists at D_2 family dopamine receptors (i.e., D_2, D_3, and D_4 receptors). L-DOPA acts as a direct dopamine receptor agonist after its conversion to dopamine. With each of these classes of drugs, an initial pharmacologic effect is achieved in the brain with the first dose, but many salient behavioral effects lag, often by many weeks. When administered with adequate dose, frequency, and chronicity, the psychostimulants produce a mixture of sensitization, tolerance, and dependence that can result in addictive behaviors (i.e., compulsive drug use despite negative consequences). Antipsychotic drugs only produce their full therapeutic effects over a period of weeks or longer. With long-term use in Parkinson's disease, L-DOPA may produce involuntary movements and psychotic symptoms, effects that can limit therapy.

For purposes of investigation, it is useful to conceptualize the acute effects of these drugs, which are mediated by their initial target proteins in the nervous system, as serving to drive neural adaptations which, with chronic or repeated drug administration, result in substantial alterations in neural function. For example, cocaine-induced increases in synaptic dopamine (resulting from its effect on the dopamine transporter) can be seen as an initiating event for longer term changes that result in sensitization, tolerance, and dependence.[1]

Adaptations to chronic drug administration are rooted in homeostatic mechanisms that permit cells to maintain their equilibrium in the face of changes in their environment. Within the brain, adaptations may occur at the level of receptors (e.g., desensitization, altered affinity, or changes in receptor number), at the level of intracellular signal transduction including alterations in gene expression, and at the level of circuitry (e.g., activation of agonist or antagonist processes that alter cell-cell communication). Changes in gene expression that occur in response to potent pharmacologic stimuli or other salient environmental or interoceptive stimuli may serve short term effector functions but also serve to prepare cells for possible repetitions of the stimulus. Adaptations to repeated or chronic drug administration, including changes in gene expression, are often complex, with varying timecourses of onset and offset.[1] This review will focus on a subset of the adaptations caused by psychostimulant drugs *in vivo* resulting in altered intracellular signaling and gene expression, and potentially contributing to behaviorally relevant alterations in synaptic function. Mechanistic investigations of these effects of psychostimulants have required the use of primary cell culture models in which dopamine and directly acting dopamine agonists have been used as stimuli. It should be noted that in addition to the effects described below, psychostimulant drugs can produce many other adaptations in neural function which differ according to the dose and pattern of administration, and which may be more or less prominent contributors to behavior depending upon the dosage regimen and the time points after the last dose investigated.

2. TRANSCRIPTIONAL REGULATION BY NEUROTRANSMITTERS AND DRUGS

While many regulated steps intervene between a gene and an active protein, it appears that extracellular signals exert their most significant effects on gene expression at the level of transcription initiation. The control of transcription depends on specific sequences of nucleotides within the regulatory regions of genes; such DNA sequences are called *cis*-regulatory elements. These sequences exert control by serving as binding sites for proteins called transcription factors. Many transcription factors bind DNA directly; others interact only indirectly via protein-protein interactions with factors that do bind DNA. DNA sequences that bind transcription factors which respond to extracellular signals are called response elements. Extracellular signals such as dopamine, which act via receptors linked to heterotrimeric G proteins, activate or repress transcription by causing reversible phosphorylation of certain transcription factors (Figure 1).

Transcription factors regulated by extracellular signals such as dopamine or glutamate, may interact directly with genes involved in the differentiated function of neurons; they may also regulate such genes indirectly by activating genes that encode other transcription factors. Genes that are activated rapidly by extracellular signals without the need for new protein synthesis are called immediate early genes (IEGs). Many IEGs encode transcription factors which may act within signaling cascades to activate downstream target genes which encode proteins that directly regulate neuronal functions.

3. REGULATION OF STRIATAL GENE EXPRESSION BY PSYCHOSTIMULANTS AND DOPAMINE

Approximately 95% of the neurons in the striatum are described morphologically as medium spiny neurons. They receive major glutamatergic inputs from the cerebral cortex and dopaminergic inputs from the substantia nigra pars compacta (SNc) and ventral tegmental area (VTA). The most extensively studied mechanisms of pharmacologically regulated gene expression in medium spiny neurons have been those mediated via dopamine D1 receptor stimulation and via D2 receptor blockade.

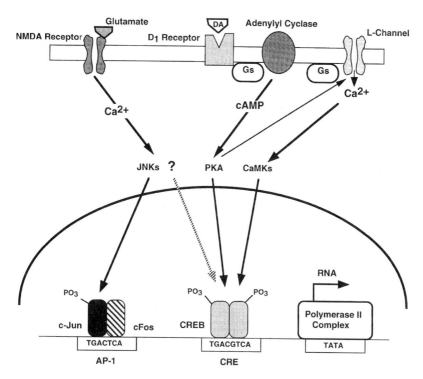

FIGURE 1.
Signaling to the nucleus by dopamine and glutamate in striatal neurons. This schematic diagram summarizes the pathways that we have investigated in striatal neurons to date. As in other cell types, mechanisms that activate the cyclic AMP pathway (such as D1 receptor stimulation in striatal neurons) or lead to Ca^{2+} entry via L-type Ca^{2+} channels lead to CREB phosphorylation and therefore activation of transcription of genes that contain cyclic AMP response elements (such as the prodynorphin and c-*fos* genes). In striatal neurons, D1-mediated CREB phosphorylation appears to require NMDA receptor activation as well; the mechanism by which NMDA receptors interact with the D1 receptor pathway is not yet clear (striped arrow). It now appears that the activation of AP-1-mediated transcription requires not only the induction of AP-1 proteins, but also activation of JNKs. In striatal neurons NMDA receptors, but not dopamine, activate JNKs.

Evidence that D1 family receptors play a necessary role in mediating the effects of psychostimulants on striatal gene expression has been based on the ability of the D1/D5 dopamine receptor antagonist, SCH23390, to block cocaine- or amphetamine-mediated induction of both IEGs and neuropeptide genes.[2-4] L-DOPA and direct D1 receptor agonists induce robust D1 receptor-mediated gene expression in the striatum following 6-hydroxydopamine lesions of the substantia nigra but not in the intact striatum. Because dopamine denervation induces D1 receptor supersensitivity such observations suggest that direct agonists are weaker stimuli than cocaine or amphetamine. There are many potential explanations for this. Dopamine produced from L-DOPA may be rapidly taken up by presynaptic dopamine neurons in the intact striatum (these neurons are destroyed by the lesion). Selective D1 agonists are not taken up but may have weaker effects than psychostimulants because the latter, as indirect dopamine agonists, lead to stimulation of D2 as well as D1 receptors. Interactions between D1 and D2 agonists have been shown to yield synergistic effects on gene expression.[5] In addition, psychostimulants can activate serotonergic neurotransmission in the striatum as well. Newly discovered classes of serotonin receptors which are expressed in the striatum are positively coupled to the adenylyl cyclase and may therefore have additive effects with D1 receptors. Of considerable interest, there is now strong evidence that D1 agonists (including psychostimulants) and D2 receptor antagonists depend on intact glutamatergic inputs to produce full regulation of striatal gene expression.[6-8] While some degree of dopamine-glutamate interaction occurs at the level of circuits, our lab has recently found that a significant component of the interaction may occur intracellularly within medium spiny neurons (see below).

3.1 Gene Regulation Mediated via D1 Dopamine Receptors

Dopamine D1 receptors are coupled via Gs and Golf to activation of the cAMP pathway[9,10] and in many cell types, including striatal neurons, to the activation of L-type Ca^{2+} channels.[11] As a result, the effects of D1 receptor stimulation include activation of the cAMP-dependent protein kinase (PKA) and likely several Ca^{2+}-dependent protein kinases of the multifunctional Ca^{2+}/calmodulin protein kinase (CaMK) family. Families of transcription factors that have been shown to be regulated via PKA and CaMKs in neurons and neuron-like cells include both constitutively expressed transcription factors, most notably the cAMP response element binding protein (CREB), and indirectly, IEG transcription factors, which include c-Fos and certain other AP-1 proteins and also the zinc finger protein, Zif268 (Egr1, NGFIa).

The major direct target of the cAMP pathway and hence of PKA is CREB and related proteins including the cyclic AMP response modulators (CREMs) and activating transcription factor-1 (ATF-1). In most cell types CREB is constitutively synthesized and is already bound to cAMP regulatory elements (CREs) as a homodimer prior to stimulation. CREB is converted into a transcriptional activator when phosphorylated on its Ser^{133} by PKA.[12,13] In contrast, some of the CREMs, which may have either activator or repressor effects, are not constitutively synthesized but instead have been shown to be inducible.[14] In addition to PKA, CaM kinases have been shown also to be capable of phosphorylating CREB.[15,16] CREB is also phosphorylated in response to other stimuli, including growth factors,[17] that activate the Ras signaling pathway, making it a potential integrator of diverse signals within neurons.

Because its promoter contains elements that bind CREB, the c-*fos* gene and several other IEGs encoding AP-1 proteins may be involved in transcriptional responses produced by cyclic AMP and Ca^{2+}, and therefore by D1 receptor stimulation. Transcriptional activation mediated by AP-1 proteins is complex, however. For example, the c-*jun* promoter does not contain a CRE (it appears to be most strongly activated by a c-Jun/ATF2 heterodimer) and more importantly, c-Jun protein, must be phosphorylated in order to fully activate transcription. Thus, as will be discussed below, induction of expression AP-1 proteins such as c-Fos and c-Jun by D1 stimulation does not necessarily entail transcriptional activation of AP-1 regulated target genes.

3.2 Gene Regulation Induced by Psychostimulants

About half of striatal medium spiny neurons in the dorsal striatum project to the globus pallidus (striatopallidal neurons), the other half to the substantia nigra pars reticulata (striatonigral neurons). Both types of medium spiny neurons utilize gamma aminobutyric acid (GABA) as their classical neurotransmitter; they differ however, in both the neuropeptides they synthesize and release and also in the types of dopamine receptor that they express. Striatopallidal neurons express the proenkephalin gene and predominantly express D2 dopamine receptors.[17a-20] Striatonigral neurons express the prodynorphin and preprotachykinin genes and predominantly express D1 receptors.[17a,18,20] The precise percentage of medium spiny neurons that co-express both D1 and D2 dopamine receptors has been a matter of controversy, but the degree of segregation appears to be biologically significant with respect to gene regulation (Figure 2). (See also Chapter 1.) Amphetamine and cocaine have been shown to induce expression of IEGs as well as of neuropeptide genes in both the dorsal and ventral striatum, predominantly in the striatonigral neurons.

3.2.1 Regulation of IEG Expression by Psychostimulants

Both amphetamine and cocaine induce expression of c-*fos* mRNA and its protein product in the striatum,[2-4] and have also been shown to induce expression of other IEG transcription factors[21] including the zinc finger transcription factor zif268 (NGFI-A, egr1).[22] In addition, both amphetamine[4] and cocaine[21,23] induce AP-1 binding activity in striatal cell extracts.

In both the dorsal and ventral striatum, psychostimulant induced IEG expression is blocked by the D1 dopamine receptor family antagonist,

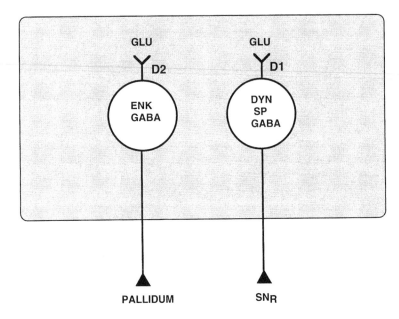

FIGURE 2.
Phenotypic characteristics of striatonigral and striatopallidal neurons. Medium spiny neurons projecting to the globus pallidus and to the substantia nigra are intermixed in equal proportions within the striatum. Both types utilize GABA as their classical neurotransmitter, but differ with respect to their peptide co-transmitters and predominant dopamine receptor type.

SCH-23390.[2-4] Consistent with this pharmacologic finding, anatomic studies demonstrate that psychostimulants induce IEGs primarily in striatonigral neurons (which express D1 dopamine receptors). By double label *in situ* hybridization, cocaine-induced c-*fos* mRNA has been shown to colocalize with preprotachykinin mRNA across postnatal developmental stages.[24] Another group using double label *in situ* hybridization[25] found a lesser degree of segregation of amphetamine-induced-*fos*, but the same overall pattern. Following amphetamine administration (5 mg/kg), c-*fos* was induced predominantly in preprotachykinin expressing neurons (77% of Fos expressing cells) rather than enkephalin expressing neurons (33%).

3.2.2 Regulation of Neuropeptide Gene Expression by Psychostimulants

In addition to inducing expression of IEGs, psychostimulants have also been shown to induce expression of preprotachykinin mRNA and prodynorphin mRNA and peptides in striatonigral neurons. Prodynorphin gene expression is induced in the intact striatum by amphetamine or cocaine and by L-dopa or D1 agonists following a 6-hydroxydopamine

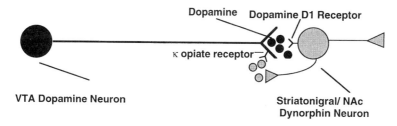

FIGURE 3.

Proposed effects of dynorphin peptides on dopamine release in striatum. Evidence from several laboratories suggests that recurrent collaterals from striatonigral neurons in the dorsal striatum and from neurons in the nucleus accumbens release GABA and dynorphin in the dorsal and ventral striatum. Release of dynorphin in the vicinity of presynaptic dopamine terminals is thought to act, via κ opioid receptors, to decrease dopamine release. Induction of dynorphin peptides by the indirect dopamine agonist psychostimulants can therefore be conceptualized as a compensatory adaptation to excessive dopamine stimulation. κ opioid receptor agonists are associated with aversive/dysphoric effects.

lesion.[17a] For example, in dorsal and ventral striatum, dynorphin mRNA and peptides have been shown to be induced significantly by repeated administrations of cocaine[25,25a,27] or methamphetamine[28] and at lower levels by a single injection of amphetamine or methamphetamine.[29] A significant increase in prodynorphin mRNA is also observed after rats self-administer cocaine.[30,31] In post-mortem studies of cocaine-dependent human drug abusers, prodynorphin, but not proenkephalin mRNA, was elevated in the striatum and nucleus accumbens.[32] In a quantitative study, chronic amphetamine (5 mg/kg for 14 days) increased prodynorphin mRNA (42% above basal) and preprotachykinin mRNA (23% above basal) but not proenkephalin or D1 or D2 receptor mRNA.[25] In contrast, in mice lacking D1 receptors as a result of homologous recombination, prodynorphin mRNA is decreased in striatal neurons.[33]

3.2.3 Regulation of Prodynorphin Gene Expression May Represent a Biologically Significant Compensatory Adaptation to Psychostimulant Administration

The regulation of prodynorphin gene expression by dopamine may represent a biologically significant homeostatic mechanism (Figure 3). Via recurrent collaterals, dynorphin peptides may feed back on dopamine terminals within the striatum to decrease dopamine release.[27,34,35] Dynorphin A and related prodynorphin gene products exhibit high affinity for kappa opiate receptors which are coupled to Go and Gi.[36,37] Kappa opiate receptors are found on the presynaptic terminals of dopamine neurons, and exert inhibitory effects on neurotransmitter release. Kappa receptors are also expressed by striatal neurons, thus the effects of dynorphin or kappa agonist drugs are complex.[34,35] Nonetheless, the overall effects of dynorphin peptides are to decrease dopamine release and possibly to antagonize the effects of dopamine on striatal neurons that co-express dopamine D1 and kappa opiate receptors. These functional neuroanatomic findings are consistent with behavioral observations. Kappa receptor agonists are associated with an aversive dysphoric syndrome in both humans[38] and rat.[39] Thus an increase in dynorphin peptides following chronic psychostimulant administration may be a compensatory adaptation to excess dopamine stimulation and decreased dynorphin in D_1 receptor knockout mice; a compensatory adaptation to diminished dopamine stimulation. By inhibiting the release of dopamine, psychostimulant-induced dynorphin peptides may contribute to the dysphoric state that occurs with the cessation of drug use and may therefore contribute to the motivational aspects of drug withdrawal.

3.2.4 Molecular Mechanisms of Striatal Prodynorphin Gene Regulation

Induction of prodynorphin gene expression by dopamine has also been shown to be dependent on D1 receptors in embryonic day 18 dissociated striatal cultures.[42] The 5′ flanking region of the rat prodynorphin gene contains three asymmetric CREs (Figure 4) which are termed DynCRE1 (−1660/−1553), DynCRE2 (−1630/−1623), and DynCRE3 (−1546/−1539). These CREs have been shown by deletion and single base mutational analysis to be required for full induction of the prodynorphin gene via cAMP pathways in CV-1 fibroblasts.[40] In addition, a noncanonical AP-1 like site (TGACAAACA; −257/−249) within the prodynorphin promoter had been previously reported to serve as a target for Fos/Jun trans-activation in NCB20 neuroblastoma cells.[41] However, we have not observed significant binding to this putative AP-1 site under either stimulated or unstimulated conditions in extracts made from striata or striatal cultures.[42]

To ensure that the prodynorphin CREs were functional in striatal neurons as well as in CV-1 cells, we performed a limited transfection analysis in primary striatal cultures. Transfection of a construct containing the 3 upstream prodynorphin CREs as well as the putative noncanonical prodynorphin AP-1 element (spanning bases -1858 to +135 with respect to the transcription start site) fused to the chloramphenicol acetyltransferase (CAT) reporter gene produced 7- to 10-fold increases in CAT activity after stimulation with dopamine (50 μM) and a phosphodiesterase inhibitor. Deletions of the two 5′ DynCREs (spanning nucleotides -1602 to +135) or all three DynCREs (spanning nucleotides −1500 to +135), but still containing the AP-1-like element dramatically reduced the ability of dopamine to induce CAT activity to

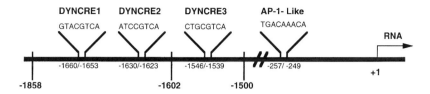

Prodynorphin Gene 5' Flanking Region

FIGURE 4.
The prodynorphin 5' flanking region. Shown are the three upstream cAMP response elements (CREs) and the putative AP-1-like element. As described in the text, the CREs have been shown to be required for dopamine induction of the prodynorphin gene in striatal neurons and to bind CREB. The AP-1-like element does not appear to bind proteins in striatal extracts with any significant affinity under conditions which give strong binding to a consensus AP-1 site (Adapted from Cole et al., 1995).

less than twofold).[42] These results agree entirely with the findings in CV-1 fibroblasts,[40] i.e., that the three upstream DynCRE elements are required for significant induction of the prodynorphin gene in response to stimuli that act via the cAMP pathway.

Having shown that they are likely to be functional in striatal neurons, we used the three dynorphin CREs (core sequences ATCCGTCA, ATCCGTCA, CTGCGTCA, respectively) as probes in electrophoretic mobility shift assays using extracts prepared from striata of rats treated with saline or amphetamine (4 mg/kg) or from E18/19 primary striatal cultures treated with vehicle, with dopamine (50 μM) together with an inhibitor of cyclic AMP phosphodiesterase, Ro20-1724 (50 μM), or forskolin (10 μM) plus Ro20-1724 (50 μM). In extracts made from either striata or primary striatal cultures we observed CREB binding to each of the three dynorphin CREs based on the patterns of competition and antibody supershift analysis. Using extracts from primary striatal cultures we observed that dopamine or forskolin also induced phosphorylation of CREB on Ser[133]. The binding of phosphorylated CREB to the prodynorphin CREs could be detected by electrophoretic mobility shift assay with an antiserum specific for phosphorylated CREB.[42]

3.2.5 Phosphorylation of CREB in Response to Dopamine Is Rapid and Dependent upon D1 Dopamine Receptor Activation

The induction of a phosphoCREB supershifted band by dopamine in the striatal cultures is blocked by pretreatment with SCH-23390, indicating that the phosphorylation of CREB in response to dopamine treatment is dependent upon D1 dopamine receptor activation. Pretreatment with the selective D2 dopamine receptor antagonist, eticlopride (50 μM), has no effect on dopamine-induced CREB phosphorylation. The selective D1 dopamine receptor agonists, SKF-82938 (50 μM) and SKF-38393 (50 µM), both induced CREB phosphorylation and this effect was not altered by eticlopride (50 µM) pretreatment. The D2/D3/D4 dopamine receptor agonist quinpirole (50 µM) did not induce CREB phosphorylation in the cultures. Overall these data indicate that D1, but not D2 family dopamine receptors are required for CREB phosphorylation in striatal culture.[42]

3.2.6 Following Chronic Amphetamine Administration, PhosphoCREB-like Immunoreactivity Is Elevated and c-fos mRNA Is Suppressed in Rat Striatum

There are low basal levels of phosphoCREB-like immunoreactivity (phosphoCREB-LI) in the rat striatum. After either a single injection of saline[42] or 12 days of daily saline injections,[42] the phosphoCREB-like immunoreactivity remains low in the striatum. PhosphoCREB-LI is markedly induced 15 min after a single acute amphetamine injection[43a] or 15 min after an amphetamine injection given after 11 days of daily saline injections.[42] Following an acute amphetamine injection (only single injections have been examined), phosphoCREB-LI rapidly returns to basal levels (approximately within an hour). However, in a chronic treatment paradigm in which 11 days of daily amphetamine injections (10 mg/kg) are followed on day 12 by a saline injection or no injection, phosphoCREB-LI is significantly elevated above control levels, although to a lesser degree than observed in acutely treated rats.[42] This upregulation of CREB phosphorylation 16 h after the final injection of amphetamine could have significant implications not only for prodynorphin, but for other cAMP-regulated genes. Because the upregulation of phosphoCREB-LI is observed whether there is a final saline injection or not, it does not represent a conditioned response to injection. It now becomes an important problem to investigate how the balance of protein kinases and phosphatases is altered by chronic amphetamine.

3.3 A Role for Glutamate: NMDA Receptor Antagonists Inhibit Acute Amphetamine-Induced IEG Expression

It has been reported that amphetamine induced-c-*fos* and *zif*268 gene expression are inhibited by NMDA receptor antagonists *in vivo*.[5-7] Similar results have been reported for cocaine. Potential explanations for these observations have generally focused on neural circuitry. NMDA-receptor blockade has been hypothesized to inhibit the release of dopamine and perhaps serotonin within the striatum or to inhibit dopamine D_1 receptor function.[43,43a,44,45] The possibility must also be considered, however, that Ca^{2+} entry via NMDA receptors might interact with the D_1 receptor-activated cyclic AMP cascade within striatal neurons to activate IEG expression. Because a number of potential substrates exist for dopamine-glutamate interactions within the circuitry controlling striatal inputs, we have studied both whole animal preparations and also dissociated embryonic day 18 (E18) neuron-enriched primary striatal cultures in which the confounding variables introduced by circuitry would be eliminated. To ensure that our results were not biased by peculiarities of the regulation of particular IEGs, (e.g., c-*fos* autoregulation), we studied the regulation of multiple IEGs.[46]

3.3.1 NMDA Receptor Antagonists Inhibit D1 Dopamine Receptor-Mediated Gene Expression in Cultured Striatal Neurons

To examine whether interactions between NMDA receptors and dopamine receptors occur only at the level of striatal circuitry or whether significant interactions occur intracellularly within striatal neurons, we used dissociated neuron-enriched primary striatal cultures that lack presynaptic dopamine and glutamate terminals. Levels of both c-*fos* and *junB* mRNA were examined in all experiments in primary striatal cultures. No significant differences were observed in the responses of these two genes. The culture medium used (D-MEM/F12) contained glutamate 2-5 mM at the time of experiments. Dopamine and other drugs were added *de novo* with each experiment.

Dopamine, and the D_1 agonists SKF-38393 and SKF-82958 all induced c-*fos* in the dissociated cultures consistent with our previous findings.[42,43a] Surprisingly, however, MK-801 (50 µM) or the competitive NMDA receptor antagonist APV (100 µM) almost completely blocked the induction of c-*fos* and c-*jun* mRNA by dopamine or the D1 agonists.[46] When cells were grown for 6 or 18 h on Ca^{2+}-free medium, the ability of dopamine to induce IEGs was also markedly reduced.[46]

Since D_1 receptor-mediated activation of protein kinase A (PKA) enhances the activity of L-type voltage sensitive Ca^{2+} channels in striatum,[11] we tested the effects of the L-type Ca^{2+} channel blockers nifedipine (20 nM to 20 µM) and verapamil (20 nM to 20 µM) on dopamine-mediated IEG induction. Neither drug inhibited dopamine mediated c-*fos* induction; indeed at higher concentrations they appeared to increase the induction.[46]

3.4 NMDA Receptor Antagonists Inhibit CREB Phosphorylation in Cultured Striatal Neurons

We have previously shown that dopamine receptor-mediated c-*fos* induction in striatal neurons is dependent upon the constitutively bound transcription factor CREB,[43a] which interacts with the c-*fos* gene at its CaRE site[47] and is activated by phosphorylation of its Ser[133]. In gel-mobility shift assays using the c-*fos* CaRE site as a probe, we observed no quantitative difference in total CREB binding following dopamine or NMDA receptor antagonists. When we examined phosphorylation of CREB, however, using a specific antiserum directed against Ser[133] phosphoCREB[48] to supershift the specific CREB-containing band, we find that treatment of cultures with MK-801 (50 µM) or APV (100 µM) decreased Ser[133] CREB phosphorylation below basal levels and blocked dopamine-induced CREB phosphorylation.[46] Thus inhibition of NMDA receptor function inhibits not only dopamine D1 receptor-induced c-*fos* gene expression, but also D1 receptor-induced CREB phosphorylation. Since we observed that NMDA receptors do not affect D1 receptor-induced cAMP levels, the interaction of the NMDA receptor and D1 pathways is likely to be downstream of the adenylyl cyclase and upstream of CREB phosphorylation.

3.5 Potential Functional Roles of AP-1 Proteins in Adaptations to Psychostimulants

Unlike CREB, which is constitutively synthesized and regulated largely by protein phosphorylation, proteins of the Fos family are induced from very low levels of expression with neuronal stimulation. For example, with acute cocaine or amphetamine administration, c-Fos is highly induced in the striatum in substance P/dynorphin expressing neurons.[24,25] With repeated stimulation, induction of c-*fos*, c-*jun*, *junB*, and *zif* 268 is markedly downregulated.[21] In contrast, certain Fos-related antigens, which may represent truncated products of the FosB gene accumulate with chronic cocaine treatment or other chronic stimuli and are expressed persistently.[49] Alterations in the composition of AP-1 complexes over time might permit cells to activate specific programs of adaptation appropriate to the strength and time-course of the stimulus.

There have been misunderstandings among some neurobiologists about the role of IEGs in general, and AP-1 proteins in particular in the regulation of neural gene expression. One misunderstanding is the idea that IEGs are a necessary intermediate in the signal-induced expression of most neural genes involved in the differentiated function of neurons. In fact, many genes involved in differentiated neural functions (e.g., genes encoding neuropeptides, proenkephalin[50] and prodynorphin,[42] and genes encoding some of the neurotrophic factors) are activated in response to neuronal depolarization or cyclic AMP via phosphorylation of CREB. In parallel, the CREB/ATF proteins also activate expression of AP-1 family IEGs[15,47,51] which can then contribute to the regulation of additional target genes.

A second misunderstanding found in the neurobiological literature is that increased expression of AP-1 proteins or increased binding of AP-1 proteins to DNA as measured by gel shift assays are often assumed to indicate increases in transcriptional activation. This assumption is unwarranted. Both c-Fos and c-Jun are phosphoproteins. Phosphorylation of certain sites within the c-Jun protein markedly increases its ability to activate transcription without affecting its ability to heterodimerize or bind DNA. For example, phosphorylation of Ser[63] or Ser[73] within the c-Jun activation domain robustly increases its ability to activate transcription, but such regulation would not be detected by measuring c-*jun* mRNA or protein or by measuring AP-1 binding by electrophoretic mobility shift assay.[52] Phosphorylation of Ser[63] and Ser[73] results from the action of Jun N terminal kinases (JNKs) which were first found to be activated by cellular stressors such as ultraviolet irradiation[53] and by interleukin-1. The degree to which Fos and Jun family members are phosphorylated at functionally significant sites by different signaling pathways in the nervous system determines whether the induction of AP-1 binding actually leads to transcriptional activation.

3.6   Evidence for Dopamine-Glutamate Interactions in Regulation of AP-1-Mediated Gene Expression in Striatal Neurons

In preliminary studies we find that in E18 primary cultures of striatal neurons, dopamine and D1 agonists induce expression of c-*fos* and also induce AP-1 binding as measured by gel-shift assay, but have no effect on JNK activation. Significantly, despite the induction of AP-1 binding, dopamine and D1 agonists have no effect on a transfected gene construct in which the reporter gene, luciferase, is driven by 4 AP-1 sites in tandem. In contrast, glutamate, acting via NMDA receptors, induces AP-1 binding and JNK activity, and markedly induces AP-1-mediated transcription (Schwarzschild et al., in preparation). Further investigation is needed, but these early results suggest that chronic psychostimulant treatments which lead to the induction of chronic FRAs and high levels of AP-1 binding in striatum[49] set the stage for substantial transcriptional effects when cells expressing the chronic FRAs are activated by NMDA receptor stimulation or other cellular activators of JNKs or other kinases acting on AP-1 proteins.

4. CONCLUSIONS

Taken together, our recent results demonstrate biologically significant interactions between dopamine and glutamate in both CREB and

AP-1-mediated transcriptional activation. Specifically, D_1 dopamine receptor-mediated gene expression in striatal neurons appears to require an interaction between two simultaneously active signaling pathways. Consistent with data from neurophysiologic approaches,[54,55] such interactions are likely to be significant in neural adaptations to psychostimulants in the intact striatum and in the development of supersensitivity in the parkinsonian striatum. Given the global actions of dopamine in the striatum, it is interesting that co-stimulation with glutamate, which has much greater specificity in its patterns of release, may be required for dopamine to exert its full transcriptional effects mediated either via the CREB or the AP-1 pathways.

ACKNOWLEDGMENT

This work was supported by PHS grants DA07134 and DA00257.

REFERENCES

1. Hyman, S. E. and Nestler, E., *The Molecular Basis of Psychiatry.* American Psychiatric Association, Washington, D.C., 1993.
2. Graybiel, A. M., Moratalla, M. R., and Robertson, H. A., Amphetamine and cocaine induce drug specific activation of the c-*fos* gene in striasome-matrix compartments and limbic subdivisions of the striatum. *Proc. Natl. Acad. Sci. U.S.A.,* 87, 6912, 1990.
3. Young, S. T., Porrino, L. J., and Iadarola, M. J., Cocaine induces striatal c-*fos*-immunoreactive proteins via dopaminergic D1 receptors. *Proc. Natl. Acad. Sci. U.S.A.,* 88, 291,1991.
4. Nguyen, T. V., Kosofsky, B., Birnbaum, R., Cohen, B. M., and Hyman, S. E., Differential expression of c-*fos* and zif 268 in rat striatum after haloperidol, clozapine, and amphetamine. *Proc. Natl. Acad. Sci. U.S.A.,* 89, 4270, 1992.
5. Gerfen, C. R., Keefe, K. A., and Gauda, E. B., D1 and D2 dopamine receptor function in striatum: Co-activation of D1- and D2-dopamine receptors on separate populations of neurons results in potentiated immediate early gene response in D1-containing neurons. *J. Neurosci.,* 15:8167, 1995.
6. Wang, J. Q., Daunais, J. B., and McGinty, J. F., NMDA receptors mediate amphetamine-induced upregulation of zif/268 and preprodynorphin mRNA expression in rat striatum. *Synapse,* 18, 343, 1994.
7. Snyder-Keller, A.M., Striatal c-*fos* induction by drugs and stress in neonatally dopamine-depleted rats given nigral transplants: importance of NMDA activation and relevance to sensitization phenomena. *Exp. Neurol.,* 113, 155, 1991.
8. Konradi, C., Cole, R. L., Senatus, P., Leveque, J. C., Pollack, A., Grossband, S. J., and Hyman, S. E., Analysis of proenkephalin second messenger-inducible enhancer in rat striatal cultures. *J. Neurochem.,* 65, 1007, 1995.
9. Kebabian, J. B. and Calne, D. B., Multiple receptors for dopamine. *Nature,* 277, 93, 1979.

10. Monsma, F. J., Mahan, L. C., Vittie, L. D., Gerfen, C. R., and Sibley, D. R., Molecular cloning and expression of a D1 dopamine receptor linked to adenylyl cyclase activation. *Proc. Natl. Acad. Sci. U.S.A.*, 87, 6723, 1990.

11. Surmeier, D. J., Bargas, J., Hemmings, H. C., Jr., Nairn, A. C., and Greengard, P., Modulation of calcium currents by a D1 dopaminergic protein kinase/phosphatase cascade in rat neostriatal neurons. *Neuron*, 14, 385, 1995.

12. Montminy, M. R. and Bilezikjian, L. M., Binding a nuclear protein to the cyclic-AMP response element of the somatostatin gene. *Nature*, 328, 175, 1987.

13. Gonzalez, G. A. and Montminy, M. R., Cyclic AMP stimulates somatostatin gene transcription by phosphorylation of CREB at serine 133. *Cell*, 59, 675, 1989.

14. Molina, C. A., Foulkes, N. S., Lalli, E., and Sassone-Corsi, P., Inducibility and negative autoregulation of CREM: an alternative promotor directs the expression of ICER and early response repressor. *Cell*, 74, 875, 1993.

15. Sheng, M., Thompson, M. A., and Greenberg, M. E., CREB: a Ca (2+)-regulated transcription factor phosphorylated by calmodulin-dependent kinases. *Science*, 252, 1427, 1993.

16. Dash, P., Karl, K. A., Colicos, M.A., C., Prywes, R., and Kandel, E. R., cAMP response element-binding protein is activated by Ca2+/Calmodulin as well as cAMP-dependent protein kinase. *Proc. Natl. Acad. Sci. U.S.A.*, 88, 5061, 1991.

17. Bonni, A., Ginty, D. D., Dudek, H., and Greenberg, M. E., Serine 133-phosphorylated CREB induces transcription via a cooperative mechanism that may confer specificity to neurotrophin signals. *Mol. Cell. Neurosci.*, 6, 168, 1995.

17a. Gerfen, C. R., Engber, T. M., Mahan, L. C., Susel, Z., Chase, T. N., Monsma, F. J., and Sibley, D. R., D1 and D2 dopamine receptor-regulated gene expression of striatonigral and striatopallidal neurons. *Science*, 250, 1429, 1990.

18. Gerfen, C. R., McGinty, J. F., and Young III, W. S., Dopamine differentially regulates dynorphin, substance P, and enkephalin expression in striatal neurons: *in situ* hybridization histochemical analysis. *J. Neurochem.*, 11, 1061, 1991.

19. Le Moine, C., Normand, E., and Guitteny, A. F. et al., Dopamine receptor gene expression by enkephalin neurons in rat forebrain. *Proc. Natl. Acad. Sci. U.S.A.*, 87, 230, 1990.

20. Yung, K. K. L., Bolam, J. P., Smith, A. D., Hersch, S. M., Liliax, B. J., and Levey, A. I., Immunocytochemical localization of D1 and D2 dopamine receptors in the basal ganglia of the rat: light and electron microscopy. *Neuroscience*, 65, 709-730.

21. Hope, B., Kosofsky, B., and Hyman, S. E., and Nestler, E., Regulation of immediate early gene expression and AP-1 binding in the rat nucleus accumbens by chronic cocaine. *Proc. Natl. Acad. Sci. U.S.A.*, 89, 5764, 1992.

22. Moratalla, R., Robertson, H. A., and Graybiel, A. M., Dynamic regulation of NGFI-A (zif268, egr1) gene expression in the striatum. *J. Neurosci.*, 12, 2609, 1992.

23. Kosofsky, B. E., Genova, L. M., and Hyman, S. E., Postnatal age defines specificity of immediate early gene induction by cocaine in developing rat brain. *J. Comp. Neurol.*, 351, 27, 1995.

24. Kosofsky, B. E., Genova, L. M., and Hyman, S. E., Substance P phenotype defines specificity of c-*fos* induction by cocaine in developing rat striatum. *J. Comp. Neurol.*, 351, 41, 1995.

25. Jaber, M., Cador, M., Dumartin, B., Normand, E., Stinus, L., and Bloch, B., Acute and chronic amphetamine treatments differently regulate neuropeptide messenger RNA levels and fos immunoreactivity in rat striatal neurons. *Neuroscience*, 64, 1041, 1995.

25a. Sivam, S. P., Cocaine selectively increases striatonigral dynorphin levels by a dopaminergic mechanism. *J. Pharmacol. Exp.*, 256, 818, 1989.

26. Smiley, P. L., Johnson, M., and Bush, L. et al., Effects of cocaine on extrapyramidal and limbic dynorphin systems. *J. Pharmacol. Exp.*, 253, 938, 1990.

27. Steiner, H. and Gerfen, C. R., Cocaine-induced c-*fos* messenger RNA is inversely related to dynorphin expression. *J. Neurosci.*, 13, 5066, 1993.

28. Li, S. J., Sivam, S. P., and McGinty, J. F. et al., Regulation of the metabolism of striatal dynorphin by the dopaminergic system. *J. Pharmacol. Exp. Ther.*, 246, 403, 1988.

29. Smith, A. J. W. and McGinty, J. F., Acute amphetamine or methamphetamine alters opioid peptide mRNA expression in rat striatum. *Mol. Brain Res.*, 21, 359, 1994.

30. Hurd, Y. L., Brown, E. E., Finlay, J. M., Fibiger, H. C., and Gerfen, C. R., Cocaine self-administration differentially alters mRNA expression of striatal peptides. *Mol. Brain Res.*, 13, 165-170, 1992.

31. Daunais, J. B., Roberts, D. C. S., and McGinty, J. F., Cocaine self-administration increases preprodynorphin, but not c-*fos* mRNA in rat striatum. *NeuroReport*, 4, 543, 1993.

32. Hurd, Y. L. and Herkenham, M., Molecular alterations in the neostriatum of human cocaine addicts. *Synapse*, 13, 357, 1993.

33. Xu, M., Moratalla, R., Gold, L. H., Hiroi, N., Koob, G. F., Graybiel, A. M., and Tonegawa, S., Dopamine D1 receptor mutant mice are deficient in striatal expression of dynorphin and in dopamine-mediated behavioral responses. *Cell*, 79, 729, 1994.

34. Krebs, M. O., Gauchy, C., Desban, M., Glowinski, J., and Kemel, M. L., Role of dynorphin and GABA in the inhibitory regulation of NMDA-induced dopamine release in striosome and matrix-enriched areas of the rat striatum. *J. Neurosci.*, 14, 2435, 1994.

35. Steiner, H. and Gerfen, C. R., Dynorphin opioid inhibition of cocaine-induced D1 dopamine receptor-mediated immediate-early gene expression in the striatum. *J. Comp. Neurol.*, 353, 200, 1995.

36. Chavkin, C., James, I. F., and Goldstein, A., Dynorphin is a specific endogenous ligand for k-opioid receptor. *Science*, 215, 413, 1982.

37. Corbett, A. D., Paterson, S. J., McKnight, A. T., Magnan, J., and Kosterlitz, H. W., Dynorphin 1-8 and dynorphin 1-9 are ligands for the kappa-subtype of opiate receptor. *Nature*, 299, 79, 1982.

38. Bals-Kubik, R., Ableitner, A., and Herz, A., Neuroanatomical sites mediating the motivational effects of opioids as mapped by the conditioned place preference paradigm in rats. *J. Pharmacol. Exp. Ther.*, 264, 489, 1993.

39. Spanagel, R., Herz, A., and Shippenberg, T. S., Opposing tonically active endogenous opioid systems modulate the mesolimbic dopaminergic pathway. *Proc. Natl. Acad. Sci. U.S.A.*, 89:2046, 1992.

40. Douglass, J., McKinzie, A. A., and Pollock, K. M., Identification of multiple DNA elements regulating basal and protein kinase: A-induces transcriptional expression of the rat prodynorphin gene. *Mol. Endocrinol.*, 8, 333, 1994.

41. Naranjo, J. R., Mellstrom, B., and Achaval, M. et al., Molecular pathways of pain: Fos/Jun-mediated activation of a noncanonical AP-1 site in the prodynorphin gene. *Neuron*, 6, 607, 1991.

42. Cole, R.L., Konradi, C., Douglass J., Hyman, S. E., Neuronal adaptation to amphetamine and dopamine: molecular mechanisms of prodynorphin gene regulation in rat striatum. *Neuron*, 14, 813, 1995.

43. Kashihara, K., Hamamura, T., Okumura, K., and Otsuki, S., Effect of MK-801 on endogenous dopamine release *in vivo*. *Brain Res.*, 528, 80, 1990.

43a. Konradi, C., Cole, R. L., Heckers, S., and Hyman, S. E., Amphetamine regulates gene expression in rat striatum via transcription factor CREB. *J. Neurosci.*, 14, 5623, 1994.

44. Krebs, M. O., Desce, J. M., Kemel, M. L., Gauchy, C., Godeheu, G., Cheramy, A., and Glowinski, J., Glutamatergic control of dopamine release in the rat striatum: evidence for presynaptic N-methyl-D-aspartate receptors on dopaminergic nerve terminals. *J. Neurochem.*, 56, 81, 1991.

45. Johnson, K. M. and Jeng, Y. J., Pharmacological evidence for N-methyl-D-aspartate receptors on nigrostriatal dopaminergic nerve terminals. *Can. J. Physiol. Pharmacol.,* 69, 1416, 1991.

46. Konradi, C., Leveque, J. C., and Hyman, S. E., Amphetamine and dopamine-induced immediate early gene expression in striatal neurons depends on postsynaptic NMDA receptors and calcium, *J. Neurosci.,* 16, 1996.

47. Sheng M., McFadden, G., and Greenberg, M. E., Membrane depolarization and calcium induce c-*fos* transcription via phosphorylation of transcription factor CREB. *Neuron,* 4, 571, 1990.

48. Ginty, D. D., Kornhauser, J. M., and Thompson, M. A. et al., Regulation of CREB phosphorylation in the suprachiasmatic nucleus by light and circadian clock. *Science,* 260, 238, 1993.

49. Hope, B. T., Nye, H. E., Kelz, M. B., Self, D. W., Iadarola, M. J., Nakabeppu, Y., Duman, R. S., and Nestler, E., Induction of a long-lasting AP-1 complex composed of altered Fos-like proteins in the brain by chronic cocaine and other chronic treatments. *Neuron,* 13, 1235, 1994.

50. Konradi, C., Kobierski, L., and Nguyen, T. V. et al., The cAMP-response-element-binding protein interacts, but Fos protein does not interact with the proenkephalin enhancer in rat striatum, *Proc. Natl. Acad. Sci. U.S.A.,* 90, 7005, 1994.

51. Sassone-Corsi, P., Visvader, J., and Ferland, L. et al., Induction of proto-oncogene fos transcription through the adenylate cyclase pathway: characterization of a cAMP-responsive element. *Genes Dev.,* 2, 529, 1988.

52. Karin, M., The regulation of AP-1 activity by mitogen-activated protein kinases. *J. Biol. Chem.,* 270, 1235, 1995.

53. Sanchez, I., Hughes, R. T., and Mayer, B. J. et al., Role of SAPK ERK kinase-1 in the stress-activated pathway regulating transcription factor c-Jun. Nature, 372, 794, 1994.

54. Cepeda, C., N.A., B., and Levine, M. S., Neuromodulatory actions of dopamine in the neostriatum are dependent upon the excitatory amino acid receptor subtypes activated, *Proc. Natl. Acad. Sci. U.S.A.,* 90, 9576, 1993.

55. Wolf, M. E., White, F. J., and Hu, X. T., MK-801 prevents alterations in the mesoaccumbens dopamine system associated with behavioral sensitization to amphetamine. *J. Neurosci.,* 14:1735-1745, 1994.

Part III

Persistent Alterations in Gene Expression Involved in Behavioral Adaptations

Chapter **6**

STRIATAL GENE EXPRESSION INVOLVED IN ADAPTIVE RESPONSES TO LEVODOPA IN PARKINSONIAN MODELS: RELEVANCE TO LONG-TERM CLINICAL EFFECTS OF LEVODOPA THERAPY

George S. Robertson

CONTENTS

0-8493-8550-4/96/$0.00+$.50
© 1996 by CRC Press, Inc.

1. INTRODUCTION

Parkinson's disease is characterized by a loss of dopamine neurons comprising the nigrostriatal pathway. Replacement therapy with the dopamine precursor, levodopa, initially provides symptomatic relief to most patients. Over time, however, an increasing proportion are disabled by motor response complications such as variations in the therapeutic response ("on-off" fluctuations) and abnormal movements that typically occur during the maximal effect of levodopa (peak dose dyskinesias). Clinical studies suggest that a progressive loss of dopamine neurons and secondary postsynaptic changes contribute to the development of these problems. Consistent with the latter proposal, acute levodopa administration to rats with unilateral 6-hydroxydopamine lesions of the nigrostriatal pathway (a classic animal model for Parkinson's disease) produces a dramatic elevation of immediate-early gene (IEG) expression in the denervated striatum. IEGs encode known transcriptional regulating factors suggesting that levodopa administration may promote long-lasting changes in the striatum that may result eventually in the development of motor response complications. In this chapter, I will review the effects of both acute and chronic alterations in dopaminergic neurotransmission on striatal IEG expression and present recent evidence suggesting that IEGs may contribute to modifications in the response to levodopa possibly by modulating neuropeptide gene expression.

2. BASIC ORGANIZATION OF NEUROPEPTIDE TRANSMITTERS AND DOPAMINE RECEPTORS IN THE STRIATUM

The predominant striatal cell type is the medium spiny projection neuron which accounts for approximately 90–95% of the neuronal population in this structure.[1] Medium spiny neurons have been classified into two subtypes based on the target region innervated by their axons; approximately half project to the substantia nigra pars reticulata while the other half project to the globus pallidus.[2,3] (See also Chapter 1). Both subtypes utilize the inhibitory neurotransmitter GABA, but each contain different combinations of neuropeptides.[4,5] Striatopallidal neurons contain enkephalin while striatonigral neurons are thought to utilize both dynorphin and substance P as neurotransmitters.[4-6] Recent evidence suggests, however, that neurotensin may be localized in both striatopallidal and striatonigral neurons.[7-11] Dopamine regulates the expression of these neuropeptides by interacting with at least two (and likely more) receptor subfamilies, termed D1 and D2.[12] Lesion studies first suggested that these receptors were located largely on distinct populations of striatal neurons, with D1 receptors residing predominantly on striatonigral neurons and D2 receptors primarily on striatopallidal neurons.[13,14] Recent *in situ* hybridization studies have confirmed this organization by demonstrating that D1 receptor and substance P mRNAs are co-expressed by striatonigral neurons whereas D2 receptor and enkephalin mRNAs are co-expressed by striatopallidal neurons[15,17] (see Chapter 1). Thus alterations in substance P and/or dynorphin expression are used as an index of alterations in the striatonigral projections whereas that in enkephalin is used as a marker of the striatopallidal neuronal activity.

3. IEG INDUCTION BY ACUTE ALTERATIONS IN DOPAMINERGIC NEUROTRANSMISSION IN A RAT MODEL OF PARKINSON'S DISEASE

Fos is the protein product of the immediate-early gene (IEG), c-*fos*, and considered to be an activity marker for some neurons.[18,19] Basal

Fos expression in the striatum is very low but is rapidly elevated by a D1-mediated mechanism after systemic administration of cocaine and d-amphetamine, stimulants which greatly enhance extracellular concentrations of dopamine.[20,21] In contrast, levodopa and directly acting D1-like receptor agonists such as SKF 38393 and CY 208-243 only weakly increase Fos-like immunoreactivity in the intact striatum.[20,23] Destruction of the nigrostriatal pathway with 6-hydroxydopamine (6-OHDA), however, endows levodopa and D1-like agonists with the ability to dramatically enhance c-*fos* expression in the denervated striatum.[20,22,24,25] These increases are completely blocked by the selective D1-like receptor antagonist, SCH 23390, indicating that activation of D1-like receptors plays an important role in levodopa- and D1-like agonist-induced c-*fos* expression.[20,25] The failure of levodopa and D1-like agonists to increase Fos-like immunoreactivity in the striatum of intact animals suggests that the development of denervation-induced supersensitivity is responsible for the large increases in c-*fos* expression produced by these compounds in the 6-OHDA-denervated striatum. This is consistent with the proposal that postsynaptic changes in the striatum may contribute to the development of long-term clinical effects of levodopa such as dyskinesias in Parkinson's disease.[26,27]

Using retrograde tracing techniques, we demonstrated that the D1-like receptor agonist, SKF 38393, elevates Fos-like immunoreactivity in striatonigral neurons[28] (Figure 1). This result is consistent with 2-deoxyglucose uptake studies showing that D1-like receptor agonists activate striatonigral neurons ipsilateral to the 6-hydroxydopamine-lesioned substantia nigra.[29] In order to determine whether SKF 38393-induced Fos-like immunoreactivity was also present in striatopallidal neurons, we combined Fos-like immunohistochemistry with the detection of proenkephalin mRNA by *in situ* hybridization histochemistry. D1-like receptor-activated Fos-like immunoreactivity was infrequently found in striatopallidal neurons identified with the proenkephalin oligonucleotide probe indicating that it is preferentially localized in striatonigral neurons.[23] This finding is consistent with autoradiographic and *in situ* hybridization studies which report that D1 receptors are expressed predominantly by striatonigral neurons.[14,15,17] Taken together, these results suggest D1-like agonist-induced c-*fos* expression may be utilized as a measure of striatonigral activation.

In contrast to levodopa and D1-like agonists, D2-like receptor agonists such as quinpirole fail to elevate Fos-like immunoreactivity in the 6-OHDA-denervated striatum.[20,23,24] Instead, D2-like receptor antagonists such as haloperidol and raclopride increase c-*fos* expression in the intact striatum[30-32] (see also Chapter 2). The D2 antagonist-induced Fos-like immunoreactivity was seen in neurons expressing proenkephalin.[23] Furthermore, neuroleptic-induced Fos-like immunoreactivity was seldom

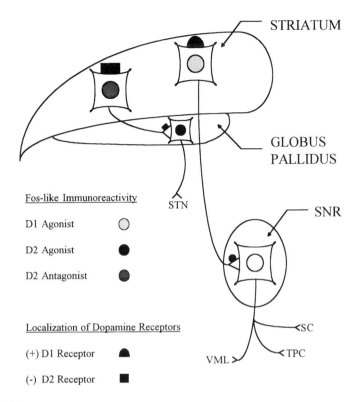

FIGURE 1.
In the 6-OHDA-denervated striatum, D1-like receptor agonists increase Fos-like immunoreactivity primarily in striatonigral neurons (lightly hatched nucleus). Selective D2-like receptor agonists enhance Fos-like expression in the ipsilateral globus pallidus (dark nucleus). D2-like receptor antagonists increase Fos-like immunoreactivity in the intact striatum, predominantly in striatopallidal neurons (darkly hatched nucleus). SC, superior colliculus; SNR, substantia nigra pars reticulata; STN, subthalamic nucleus; TPC, pedunculopontine nucleus; and VML, ventromedial and ventrolateral thalamic nuclei.

found in striatonigral neurons retrogradely labeled with Fluoro-Gold from the substantia nigra pars reticulata.[23] These findings indicate that D2-like receptor antagonists elevate c-*fos* expression primarily in striatopallidal neurons. These results are consistent with neurochemical and neurophysiological studies showing that dopamine inhibits striatopallidal activity[15,33,34] (Figure 1).

Although the D2-like agonist, quinpirole, fails to induce c-*fos* in the dopamine denervated striatum, it elevates Fos-like immunoreactivity in the ipsilateral globus pallidus (Figure 1). Electrophysiological studies have reported that D2-like receptor activation increases the activity of pallidal neurons and that 6-OHDA lesions of the nigrostriatal pathway potentiate this increase.[34] The increase in Fos-like immunoreactivity in the globus pallidus ipsilateral to the 6-OHDA-denervated striatum is consistent with the enhanced ability of quinpirole to activate pallidal neurons. The small amount of D2 receptor mRNA in the globus pallidus[35,36] suggests that the excitatory effects of quinpirole on pallidal neurons may be indirect. Given that D2 receptors reside on striatopallidal neurons and function to inhibit these neurons, it is possible that this pathway is involved in the stimulatory actions of quinpirole on pallidal activity. By inhibiting striatopallidal neurons, quinpirole would presumably decrease the release of such inhibitory neurotransmitters as GABA and enkephalin from striatopallidal terminals resulting in a disinhibition of pallidal neurons. The ability of quinpirole to elevate c-*fos* expression in the globus pallidus after 6-OHDA lesions may therefore be a reflection of D2 receptor supersensitivity in the striatum rather than the globus pallidus. If this is the case, D2-like receptor-mediated increases in c-*fos* expression in the globus pallidus may serve as an index of both pallidal and striatopallidal activity.

4. DIFFERENTIAL BEHAVIORAL AND BIOCHEMICAL EFFECTS OF CONTINUOUS AND INTERMITTENT LEVODOPA ADMINISTRATION

4.1 Effects of Levodopa Priming on Circling Induced by D1 and D2 Dopamine Agonists in 6-OHDA-Lesioned Rats

The fact that switching patients from oral, intermittent levodopa therapy to continuous levodopa administration diminishes motor response fluctuations suggests that the dosing regimen plays an important role in altering the basal ganglia neuronal activity. This has led to investigation of the role dose scheduling plays in the behavioral responsiveness to dopamine agonists.[26,37] For instance, the effects of intermittent and continuous levodopa on rotational behavior induced by selective D1- and D2-like agonists have been studied in the 6-OHDA-lesioned rat model.[39] Chronic administration of levodopa by continuous infusion (osmotic minipump for 19 days with a 3-day washout) enhanced the rotational response to the selective D2-like agonist quinpirole but had no effect on rotational behavior-induced by the selective D1-like receptor agonist SKF 38393.[39] These results suggest that continuous levodopa preserves the ability of SKF 38393 to activate striatonigral neurons but increases the inhibitory action of quinpirole on striatopallidal neurons. In contrast, chronic administration of levodopa by intermittent injection (single daily injection for 19 days with a 3-day washout) has dramatically different effects on the rotational response to selective agonists. Quinpirole-induced circling is markedly enhanced while SKF 38393-induced turning is greatly diminished, suggesting that intermittent levodopa greatly increases the inhibitory action of quinpirole on striatopallidal neurons but eliminates the ability of SKF 38393 to activate striatonigral neurons.[39] Given that patients with Parkinson's disease are more likely to suffer motor response complications with intermittent rather than continuous levodopa therapy, these findings may have relevance for the development of such problems. For instance, a reduction in striatonigral responsiveness to D1 receptor stimulation may contribute to variations, or even loss, of the therapeutic response to levodopa. On the other hand, excessive D2 receptor-mediated inhibition of striatopallidal neurons may play a role in the production of levodopa-induced dyskinesias.[39]

4.2   Effects of Chronic Administration of Dopaminergic Agonists on Neuropeptide Concentrations: Importance of Dose Scheduling

The dramatically different effects of intermittent and continuous levodopa administration on D1 and D2 receptor-mediated turning in 6-OHDA-lesioned rats led to studies of the effects of the levodopa dosing regimens on striatal neuropeptide systems. Destruction of the nigrostriatal pathway by 6-OHDA injections into the medial forebrain bundle elevates enkephalin levels in striatopallidal neurons and decreases dynorphin and substance P concentrations in striatonigral neurons.[40,41] Chronic administration of levodopa fails to restore the normal balance between striatopallidal and striatonigral neuropeptides, but rather creates a new equilibrium that depends on the constancy with which it is delivered. Continuous administration of levodopa does not reverse the 6-OHDA-induced reductions in substance P and dynorphin levels, whereas the increase in enkephalin levels produced by dopaminergic deafferentation are enhanced.[41] In contrast, intermittent levodopa administration reverses the reduction in substance P levels produced by 6-OHDA lesions but fails to alter enkephalin levels.[41] However, the most dramatic changes occur with striatonigral dynorphin concentrations in the 6-OHDA-lesioned side where intermittent levodopa produces over a ten-fold elevation.[42] Alterations in neuropeptide levels produced by dopaminergic denervation and intermittent administration of levodopa are accompanied by parallel shifts in mRNAs encoding these peptides and may reflect altered synthesis.[15,43,44] These data demonstrate that continuous and intermittent levodopa administration differentially affects striatal output, a phenomenon that may underlie differential responses to acute D1 and D2-like agonists after levodopa priming as discussed above (Section 4.1).

5. LONG-TERM INDUCTION OF IEGs BY DOPAMINERGIC AGENTS: EFFECTS ON STRIATAL NEUROPEPTIDE GENE EXPRESSION

In addition to being putative activity markers, IEG products function as transcriptional regulating factors that couple extracellular signals to alterations in cellular phenotype by regulating the expression of specific target genes.[45,46] In the case of proteins encoded by c-*fos* and c-*jun*, this process entails the dimerization of Fos and Jun via a heptad repeat of five leucine residues common to both proteins forming a complex that binds to a nucleotide sequence motif known as the AP-1 site.[47-50] AP-1-like binding sites have been identified in the promoters of genes encoding the neuropeptides enkephalin, dynorphin, cholecystokinin, neurotensin and substance P suggesting that their expression may be regulated by IEGs.[51-55] Indeed, transient co-transfection studies indicate that Fos and Jun can enhance proenkephalin and prodynorphin expression.[51,54] Similarly, we and others recently demonstrated that inhibition of haloperidol-induced c-*fos* expression by intrastriatal injection of an antisense phosphorothioate oligodeoxynucleotide to this IEG attenuates the subsequent increase in proneurotensin mRNA.[11,56] These data indicate that c-*fos* participates in those intracellular events which mediate changes in neuropeptide gene expression that occur after acute haloperidol administration. However, it does not appear that c-*fos* mediates changes in neuropeptide gene expression produced by chronic alterations in dopaminergic neurotransmission[57] (see also Chapter 2). For example, chronic administration of dopaminergic stimulants that acutely increase c-*fos* expression leads to a rapid loss in the ability of these compounds to elevate c-*fos* expression in the striatum.[58-60] Despite this desensitization, AP-1 binding remains elevated suggesting that another member of the *fos* family is responsible for the maintenance of transcriptional changes initiated by Fos.[58] Consistent with this proposal, repeated administration of the mixed D1/D2 receptor agonist, apomorphine, to 6-OHDA-lesioned rats or of cocaine (an indirect dopamine agonist) to normal animals, produces a persistent elevation of Fos-like immunoreactivity, detected with an antibody that recognizes all known members of the Fos family, in the striatum.[61-63] Western blotting and gel shift experiments indicate that a 35-kDa Fos-related antigen is at least partly responsible for the prolonged increase in AP-1 binding produced by chronic apomorphine or cocaine administration.[62,63] Moreover, this Fos-related antigen is expressed in striatonigral neurons suggesting that it may contribute to the facilitatory effects of chronic apomorphine treatment on dynorphin and substance P expression.[61]

One likely candidate contributing to the long-lasting elevation in AP-1 activity is the truncated form of FosB (ΔFosB) which is is approximately 35 kDa in size and displays prolonged induction kinetics.[63-65] Two different forms of *fosB* mRNA are generated by alternative splicing of the transcript from a single *fosB* gene.[64-68] The longer transcript (*fosB*) encodes a protein 338 amino acids in length called FosB while the shorter transcript (Δ*fosB*) encodes a truncated form of FosB known as ΔFosB. Δ*fosB* mRNA is produced by deletion of 140 bases from the *fosB* transcript. This deletion shifts the reading frame by a single base, creating the stop codon TGA. As a result, ΔFosB is only 237 amino acids long and lacks the last 101 amino acids found in FosB. Present within this truncated region is the heptoproline sequence (amino acids 257-263) that functions as an activation domain in FosB. ΔFosB is therefore a much weaker transcriptional activating factor than FosB. Nevertheless, ΔFosB activates an artificial promotor construct containing multiple AP-1 binding sites.[65,67] This quality may enable ΔFosB/Jun dimers to activate such neuropeptide genes as prodynorphin and proneurotensin which contain several AP-1-like consensus sites.[53-54]

We have recently examined the effects of chronic alterations in dopaminergic neurotransmission on expression of this protein in striatonigral and striatopallidal neurons. Briefly, our studies indicate that dopaminergic-denervation and chronic haloperidol administration produce a persistent elevation of ΔFosB expression in striatopallidal neurons.[69] In contrast, repeated administration of the D1-like receptor agonist, CY 208-243, (1 mg/kg, injected subcutaneously twice daily for 5 days) to 6-OHDA-lesioned rats generated a prolonged enhancement of ΔFosB levels in the deafferented striatum that was present principally in striatonigral neurons.[69] These results suggest that ΔFosB may be involved in the gene signaling events that mediate the long-lasting effects of chronic alterations in dopaminergic neurotransmission on striatal neuropeptide gene expression.

6. DOPAMINERGIC REGULATION OF ΔFosB EXPRESSION IN MONKEYS RENDERED PARKINSONIAN WITH MPTP

6.1   MPTP Model of Parkinson's Disease

The neurotoxin MPTP (1-methyl-4-phenyl-1,2,3,6-tetrahydropyridine) selectively destroys midbrain dopamine neurons in both humans and monkeys producing a syndrome that closely resembles Parkinson's disease.[70-72] Like humans afflicted with Parkinson's disease, MPTP-treated monkeys develop akinesia, bradykinesia, rigidity, stooped posture and tremor. Furthermore, dyskinesias are also produced in MPTP-lesioned monkeys by repeated administration of levodopa or D1-like receptor agonists.[73] These similarities indicate that the MPTP-lesioned monkey is an excellent model of Parkinson's disease in humans.

6.2 Effects of Chronic Administration of Dopaminergic Agonists on ΔFosB Expression in MPTP-Lesioned Monkeys

Given that repeated administration of the D1-like agonist CY 204-243 produces a profound elevation of ΔFosB-like protein(s) in the denervated striatum of 6-OHDA-lesioned rats, we have examined the effects of chronic administration of selective D1- and D2-like dopamine receptor agonists on ΔFosB-like protein expression in the caudate nuclei of monkeys rendered parkinsonian by treatment with MPTP. MPTP-treated monkeys were permitted a 1 month drug-free recovery period before being killed or treated chronically with a selective D1- or D2-like receptor agonist for about 1 month. As was the case with 6-OHDA-lesioned rats, the caudate nuclei of MPTP-lesioned monkeys contained elevated levels of ΔFosB-like protein(s) (43 and 45 kDa).[69] Also similar to results obtained with 6-OHDA-lesioned rats, chronic administration of the selective D1-like agonist SKF 82958 produced a further elevation of ΔFosB-like protein(s) in MPTP-denervated caudate nuclei. Since these monkeys were killed 4 days after the cessation of agonist administration, SKF 82958-induced increases in ΔFosB-like immunoreactivity appear to be long lasting. In addition, SKF 82958-induced increases in ΔFosB expression were only observed in the two monkeys that became dyskinetic. A third which received similar treatment but failed to become dyskinetic did not display enhanced ΔFosB levels in the caudate nucleus. Because chronic administration of the D1-like agonist CY 208-243 to 6-OHDA-lesioned rats elevated ΔFosB-like immunoreactivity in striatonigral neurons, these findings suggest that increased ΔFosB expression, perhaps in striatonigral neurons, is associated with the development of dyskinesia. In contrast, chronic administration of the long-acting D2-like agonist carbergoline which alleviated parkinsonian scores without producing dyskinesia reversed the elevation of ΔFosB-like protein(s) by MPTP-treatment. Our retrograde tract tracing studies in 6-OHDA-lesioned rats indicate that chronic dopaminergic denervation elevates ΔFosB-like protein(s) in striatopallidal neurons. The vast majority of striatal D2 receptors are thought to be located on striatopallidal neurons which are tonically inhibited by activation of this dopamine receptor subtype.[15,23,74] The loss of striatal dopamine subsequent to destruction of the nigrostriatal pathway would therefore eliminate D2 receptor-mediated inhibition of striatopallidal neurons. The resultant overactivity of striatopallidal neurons has been proposed to contribute to the hypokinetic state characteristic of Parkinson's disease.[75] Accordingly, chronic administration of D2-like receptor agonists such as carbergoline is thought to restore normal movement by reinstating the D2 receptor-mediated inhibition of striatopallidal neurons. Consequently, carbergoline may also reverse the elevation of ΔFosB-like protein(s) in striatopallidal neurons after dopaminergic denervation by normalizing D2 receptor-mediated signal transduction in these neurons.

7. ΔFosB PARTICIPATES IN MEDIATION OF PRIMING

Administration of dopamine receptor agonists to 6-OHDA-lesioned rats produce changes in the denervated striatum that enable a subsequent injection to elicit more vigorous circling[76] (see also Section 5). This phenomenon is termed priming and is considered to be a robust behavioral model of synaptic plasticity in the striatum. Among those dopaminergic agonists which produce priming, levodopa is the most potent (personal communication, Dr. Micaella Morelli). This suggests that determining the nature of the alterations in striatal gene expression mediating priming may lead to insights into the molecular basis for levodopa-induced dyskinesias. The strong correlation between D1-like agonist-induced ΔFosB expression and dyskinesias led us to propose that this transcription regulator may be involved in cellular mechanisms leading to dyskinesia. Consistent with this hypothesis, we recently demonstrated that ΔFosB plays an important role in priming. Thus levodopa-induced priming and ΔFosB expression show an excellent temporal correlation.[77,78] Next, we used the antisense approach (see Chapters 3 and 8) to establish a link between ΔFosB expression and priming. The question asked was: Does blockade of ΔFosB expression by central injection of antisense DNA block apomorphine-induced priming? Indeed, inhibition of apomorphine-induced ΔFosB expression by intrastriatal administration of ΔFosB antisense oligomer significantly reduced the ability of this dopamine agonist to enhance the circling response to a subsequent injection of the D1-selective agonist, SKF 38393. The antisense oligomer did not reduce apomorphine-induced expression of other IEGs, Fos, Jun, JunB or NGFI-A. Thus the antisense oligomer administration reduced apomorphine-induced priming by selectively blocking the synthesis of FosB and/or ΔFosB. Taken together, these data indicate that ΔFosB plays an important role in dopamine-receptor-mediated priming and suggests that this IEG may be involved in levodopa-induced dyskinesia.

8. POTENTIAL NEUROPEPTIDE GENE TARGETS OF CHRONICALLY INDUCED IEGs IN THE NEOSTRIATUM

It is unknown whether ΔFosB directly elevates transcription of target neuropeptide genes *in vivo*. As mentioned previously, ΔFosB is an inhibitory component of AP-1 that reduces the transactivating activity of Fos/Jun, presumably due to the absence of the C terminus transactivation domain found in FosB which has the potential to interact with TATA binding proteins.[64] Nevertheless, recent experimentation has demonstrated that while ΔFosB lacks transcriptional activity it can increase mRNA stability.[66] Thus, a persistent elevation of ΔFosB levels may contribute to the development of dyskinesias by stabilizing mRNAs which encode neuropeptide transmitters.

Chronic D1-like agonist administration elevates levels of the neuropeptide transmitters dynorphin, substance P and neurotensin in striatonigral neurons of the dopaminergically deafferented striatum.[8-10,15,79,80] Because these neuropeptides influence the activity of several nigrostriatal dopamine systems,[81,82] increases in their levels might create activity imbalances between opposing basal ganglia circuits which promote dyskinesia.[75] AP-1-like sites have been identified in preprotachykinin, neurotensin/neuromedin N and prodynorphin genes,[52-54] suggesting that their expression may be regulated by members of the *fos* and *jun* families. A potential role for AP-1 in the dopaminergic regulation of preprotachykinin gene expression has yet to be established. However, the AP-1 binding site within the neurotensin/neuromedin N promoter contributes to the inducibility of this gene by nerve growth factor in PC-12 cells.[53] Moreover, Fos is necessary for acute increases in striatal neurotensin/neuromedin N expression produced by a single administration of the D2-like receptor antagonist haloperidol.[11,56] These findings suggest that prolonged increases in AP-1 binding might mediate the elevation of neurotensin levels in the 6-OHDA-denervated striatum that have been reported to occur after chronic D1-like receptor activation.

Sustained elevation in ΔFosB expression is seen in dynorphin and substance P containing neurons in the dopamine denervated striatum.[61] Additionally, a 35-kDa Fos-related antigen appears to contribute to the long-term induction of striatal dynorphin expression in 6-OHDA-lesioned rats.[83] Finally, as discussed above (Section 4.2), after chronic intermittent levodopa treatment, the most dramatic changes are seen in the striatal dynorphin system.[42] These data along with the role of dynorphin in regulating basal ganglia functions discussed below suggest that ΔFosB-induced regulation of prodynorphin mRNA expression may underlie priming effects of levodopa.

Dynorphin interacts primarily with the kappa subtype of opioid receptor which is located in a variety of central nervous system structures including the striatum and substantia nigra.[84-86] Within the striatum, kappa receptors appear to be associated with both presynatic and postsynaptic elements, however, their precise cellular localization is still uncertain. D1 receptor stimulation increases extracellular concentration of dynorphin in the striatum indicating that this neuropeptide can be released from the collaterals of striatonigral neurons.[42] Recent evidence suggests that dynorphin released from recurrent collaterals may act to blunt the response of striatonigral neurons to dopamine input.[87,88] Indeed, systemic administration of the kappa receptor agonist spiradoline attenuates D1 receptor agonist-induced turning in 6-OHDA-lesioned rats.[41] These studies suggest that dynorphin, acting through kappa receptors in the striatum, negatively modulates the excitatory actions of D1 receptor stimulation on striatonigral activity. Elevated dynorphin stores in the striatum, leading to increased dynorphin release and kappa receptor activation, may therefore account for the reduction in D1 agonist-induced turning produced by chronic intermittent levodopa administration. Moreover, by opposing the facilitatory actions of D1 receptor-mediated neurotransmission on striatonigral activity, excessive increases in striatal dynorphin levels might contribute to variations in the therapeutic response ("on-off" fluctuations) that occur with chronic levodopa administration in parkinsonian patients.

Within the substantia nigra, kappa binding is located primarily on neurons of the pars reticulata which receive dynorphinergic input by way of the striatonigral pathway.[85,86] *In vivo* microdialysis studies have demonstrated that, like the striatum, acute D1 receptor activation enhances dynorphin release in the substantia nigra.[42] However, in contrast to the inhibitory effects of striatal dynorphin on locomotion, behavioral studies indicate that dynorphin release in the substantia nigra pars reticulata acts to facilitate movement. When administered into the substantia nigra, the specific kappa receptor agonist U-50,488H produces dose-dependent contralateral circling, probably due to direct inhibition of reticulata neurons.[89,90] Similarly, microinjections of dynorphin into the ipsilateral substantia nigra pars reticulata of 6-OHDA-lesioned rats promote contralateral rotation.[91] Accordingly, treatments which elevate striatonigral dynorphin levels in 6-OHDA-lesioned rats might be expected to enhance, rather than diminish, the rotational response to D1 receptor stimulation. Indeed, repeated administration of D1-like receptor agonists to 6-OHDA-lesioned rats will produce a progressive increase in circling if the interval between successive injections is sufficient in length.[77,78,92] It is tempting to speculate that a selective enhancement of nigral dynorphin release may mediate this sensitization phenomenon and perhaps also the development of levodopa-induced dyskinesias. By contrast, reducing the time between successive D1-like agonist injections beyond a critical point has the opposite effect producing a desensitization of D1 receptor-mediated rotation, a situation similar to that observed for D1-like agonist-induced circling after chronic intermittent levodopa administration. As proposed earlier, these behavioral changes may be mediated by increased release of dynorphin from the recurrent collaterals of striatonigral neurons.

In summary, chronic levodopa administration may initially elevate dynorphin levels in the substantia nigra pars reticulata giving rise to excessive, abnormal movements (peak dose dyskinesias). However, as treatment continues, dynorphin concentrations in the striatum may be elevated to the point where its release from recurrent collaterals shuts down striatonigral activity resulting in fluctuations, or even loss, of the therapeutic response.

The functional role of dynorphin in the basal ganglia indicates that levodopa priming-induced sustained increase in ∆FosB expression may alter the activity of dynorphin neurons and lead to clinical complications (dyskinesia, "on/off") of levodopa therapy. It is important to note however, recent studies suggest that three cAMP response elements (CREs), rather than the previously described noncanonical AP-1 site, may be critical for dopamine induction of the prodynorphin gene in striatal neurons[93] (see also Chapter 5). Thus the role of dynorphin (if any) in chronic levodopa effects remains to be tested.

REFERENCES

1. Kemp, J.M. and Powell, T.P.S., The structure of the caudate nucleus of the cat: light and electron microscopy, *Phil. Trans. R. Soc. London* 262, 383, 1971.
2. Beckstead, R.M. and Cruz, C.J., Striatal axons to the globus pallidus, entopeduncular nucleus and substantia nigra come mainly from separate cell populations in cat, *Neuroscience* 19, 147, 1986.
3. Kawaguchi, Y., Wilson, C.J. and Emson, P.C., Projection subtypes of rat neostriatal matrix cells revealed by intracellular injection of biocytin, *J. Neurosci.* 10, 3421, 1990.
4. Anderson, K.D. and Reiner, A., Extensive co-occurrence of substance P and dynorphin in striatal projection neurons: an evolutionarily conserved feature of basal ganglia organization, *J. Comp. Neurol.* 295, 339, 1990.
5. Beckstead, R.M. and Kersey, K.S., Immunohistochemical demonstration of different substance P-, met-enkephalin-, and glutamic-acid-decarboxylase-containing cell body and axon distributions in the corpus striatum of the cat, *J. Comp. Neurol.* 232, 481, 1985.
6. Gerfen, C.R. and Young, W.S. III, Distribution of striatonigral and striatopallidal peptidergic neurons in both patch and matrix compartments: an *in situ* hybridization histochemistry and fluorescent retrograde tracing study, *Brain Res.* 460, 161, 1988.

7. Fuxe, K., Von Euler, G., Agnati, L.F., Merlo Pich, E., O'Connor, W.T., Tanganelli, S., Li, X.M., Tinner, B., Cintra, A., Carani, A. and Benfenati, F., Intramembrane receptors and dopamine D2 receptors as a major mechanism for the neuroleptic-like action of neurotensin, *Ann. N.Y. Acad. Sci.* 688, 186, 1992.

8. Castel, M.N., Morino, P. and Hökfelt, T., Modulation of the neurotensin striatonigral pathway by D1 receptors, *NeuroReport* 5, 281, 1993a.

9. Castel, M.N., Morino, P., Frey, P., Terenius, L. and Hökfelt, T., Immunohistochemical evidence for a neurotensin striatonigral pathway in the rat brain, *Neuroscience* 55, 833, 1993b.

10. Castel, M.N., Morino, P., Dagerlind, A. and Hökfelt, T., Up-regulation of neurotensin mRNA in the rat striatum after acute methamphetamine treatment, *Eur. J. Neurosci.* 6, 646, 1994.

11. Robertson, G.S., Tetzlaff, W., Bedard, A., St-Jean, M. and Wigle, N., c-*fos* mediates antipsychotic-induced neurotensin gene expression in the rodent striatum, *Neuroscience*, 1995.

12. Kebabian, J.W. and Calne, D.B., Multiple receptors for dopamine, *Nature* 277, 93, 1979.

13. Harrison, M.B., Wiley, R.G. and Wooten, G.F., Selective localization of striatal D1 receptors to striatonigral neurons, *Brain Res.* 528, 317, 1990.

14. Harrison, M.B., Wiley, R.G. and Wooten, G.F., Changes in D2 but not D1 receptor binding in the striatum following a selective lesion of striatopallidal neurons, *Brain Res.* 590, 305, 1992.

15. Gerfen, C.R., Engber, T.M., Mahan, L.C., Susel, Z., Chase, T.N., Monsma, F.J., Jr. and Sibley, D.R., D_1 and D_2 dopamine receptor-regulated gene expression of striatonigral and striatopallidal neurons, *Science* 250, 1429, 1990.

16. Gerfen, C.R., McGinty, J.F. and Young, W.S. III, Dopamine differentially regulates dynorphin, substance P, and enkephalin expression in striatal neurons: *in situ* hybridization histochemical analysis, *J. Neurosci.* 11, 1016, 1991.

17. Le Moine, C., Normad, E. and Bloch, B., Phenotypical characterization of rat striatal neurons expressing the D1 dopamine receptor gene, *Proc. Natl. Acad. Sci. U.S.A.* 88, 4205, 1991.

18. Sagar, S.M., Sharp, F.R. and Curran, T., Expression of c-*fos* protein in brain: metabolic mapping at the cellular level, *Science* 240, 1328, 1988.

19. Dragunow, M. and Faull, R., The use of c-*fos* as a metabolic marker in neuronal pathway tracing, *J. Neurosci. Methods* 29, 261, 1989.

20. Robertson, H.A., Peterson, M.R., Murphy, K. and Robertson, G.S., D1-dopamine receptor agonists selectively activate c-*fos* independent of rotational behaviour, *Brain Res.* 503, 346, 1989b.

21. Graybiel, A.M., Moratalla, R. and Robertson, H.A., Amphetamine and cocaine induce drug specific activation of the c-*fos* gene in striosome-matrix compartments and limbic subdivisions of the striatum, *Proc. Natl. Acad. Sci. U.S.A.* 87, 6912, 1990.

22. Robertson, G.S., Herrera, D.G., Dragunow, M. and Robertson, H.A., L-Dopa activates c-*fos* in the striatum ipsilateral to a 6-hydroxydopamine lesion of the substantia nigra, *Eur. J. Pharmacol.* 159, 99, 1989.

23. Robertson, G.S., Vincent, S.R. and Fibiger, H.C., D1 and D2 dopamine receptors differentially regulate c-*fos* expression in striatonigral and striatopallidal neurons, *Neuroscience* 49, 285, 1992.

24. Paul, M.L., Graybiel, A.M., David, J.C. and Robertson, H.A., D1-like and D2-like dopamine receptors synergistically activates rotation and c-*fos* expression in the dopamine-depleted striatum in a rat model of Parkinson's disease, *J. Neurosci.* 12, 3729, 1992.

25. Morelli, M., Cozzolino, A., Pinna, A., Fenu, S., Garau, L. and Di Chiara, G., L-dopa stimulates c-*fos* expression in dopamine denervated striatum by combined activation of D-1 and D-2 receptors, *Brain Res.* 623, 334, 1993.

26. Chase, T.N., Mouradian, M.M. and Engber, T.M., Motor complications and the function of striatal efferent systems, *Neurology* 43 (Suppl. 6), S23, 1993.

27. Obseo, J.A., Grandas, F., Herrero, M.T. and Horowski, R., The role of pulsatile vs. continuous dopamine receptor stimulation for functional recovery in Parkinson's disease, *Eur. J. Neurosci.* 6, 889, 1994.

28. Robertson, G.S., Vincent, S.R. and Fibiger, H.B., Striatonigral projection neurons contain D1 dopamine receptor-activated c-*fos*, *Brain Res.* 523, 288, 1990.

29. Trugman, J.M. and Wooten, G.F., Selective D1 and D2 dopamine agonists differentially alter basal ganglia glucose utilization in rats with unilateral 6-hydroxydopamine substantia nigra lesions, *J. Neurosci.* 7, 2927, 1987.

30. Dragunow, M., Robertson, G.S., Faull, R.L.M., Robertson, H.A. and Jansen, K., D2 dopamine receptor antagonists induce Fos and related in rat striatal neurons, *Neuroscience* 37, 287, 1990.

31. Deutch, A.Y., Lee, M.C. and Iadarola, M.J., Regionally specific effects of atypical antipsychotic drugs on striatal Fos expression: the nucleus accumbens shell as a locus of antipsychotic action, *Mol. Cell. Neurosci.* 3, 332, 1992.

32. Robertson, G.S. and Fibiger, H.C., Neuroleptics increase c-*fos* expression in the forebrain: contrasting effects of haloperidol and clozapine, *Neuroscience* 46, 315, 1992.

33. Pan, H.S., Penney, J.B. and Young, A.B., γ-Aminobutyric acid and benzodiazepine receptor changes induced by unilateral 6-hydroxydopamine lesions of the medial forebrain bundle, *J. Neurochem.* 45, 1396, 1985.

34. Carlson, J., Bergstrom, D.A., Demo, S. and Walters, J.R., Nigrostriatal lesions alters neurophysiological responses to selective and nonselective D-1 and D-2 dopamine agonists in rat globus pallidus, *Synapse* 5, 83, 1990.

35. Meador-Woodruff, J.H., Mansour, A., Bunzow, J.R., Van Tol, H.H.M., Watson, S.J., Jr. and Civelli, O, Distribution of D2 dopamine receptor mRNA in rat brain, *Proc. Natl. Acad. Sci. U.S.A.* 86, 7625, 1989.

36. Mengod, G., Martinez-Mir, I., Vilaro, M.T. and Palacios, J.M., Localization of the mRNA for the dopamine D2 receptor in the rat brain by *in situ* hybridization histochemistry, *Proc. Natl. Acad. Sci. U.S.A.* 86, 8560, 1989.

37. Shoulson, I., Glaubieger, G.A. and Chase, T.N., "On-off" response: clinical and biochemical correlations during oral and intravenous levodopa administration in parkinsonian patients, *Neurology* 25, 1144, 1975.

38. Engber, T.M., Susel, Z., Juncos, J.L. and Chase, T.N., Continuous and intermittent levodopa differentially affect rotation induced by D-1 and D-2 dopamine agonists, *Eur. J. Pharmacol.* 168, 291, 1989.

39. Crossman, A.R., A hypothesis on the pathophysiological mechanisms that underlie levodopa or dopamine agonist-induced dyskinesia in Parkinson's disease: implications for future strategies in treatment, *Movement Disorders* 5, 100, 1990.

40. Hong, J.S., Yang, H.-Y.T., Fratta, W. and Costa, E., Rat striatal methionine-enkephalin content after chronic treatment with cataleptogenic and noncataleptogenic drugs, *J. Pharmacol. Exp. Ther.* 205, 141, 1978.

41. Engber, T.M., Susel, Z., Kuo, S., Gerfen, C.R. and Chase, T.N., Levodopa replacement therapy alters enzyme activities in striatum and neuropeptide content in striatal output regions of 6-hydroxydopamine lesioned rats, *Brain Res.* 552, 113, 1991.

42. You, Z.B., Herra-Marshitz, M., Goiny, N.M., O'Connor, W.T., Ungerstedt, U. and Terenius, L., The striatonigral dynorphin pathway of the rat studied with *in vivo* microdialysis-II. Effects of dopamine D1 and D2 receptor agonists, *Neuroscience* 63, 427, 1994.

43. Young, W.S., III, Bonner, T.I. and Brann, M.R., Mesencephalic dopaminergic neurons regulate the expression of neuropeptide mRNAs in the rat forebrain, *Proc. Natl. Acad. Sci. U.S.A.* 83: 9827, 1986.

44. Le Moine, C., Normand, E., Guitteny, A.F., Fouque, B., Teoule, R. and Bloch, B., Dopamine receptor gene expression by enkephalin neurons in rat forebrain, *Proc. Natl. Acad. Sci. U.S.A.* 87, 230, 1990.

45. Sheng, M. and Greenberg, M.E., The regulation and function of c-*fos* and other immediate-early genes in the nervous system, *Neuron* 4, 477, 1990.

46. Morgan, J.I. and Curran, T., Stimulus-transcription coupling in the nervous system: involvement of the inducible proto-oncogenes *fos* and *jun*, *Annu. Rev. Neurosci.* 14, 421.

47. Gentz, R., Rauscher, F.J. III, Abate, C. and Curran, T, Parallel association of Fos and Jun leucine zippers juxtaposes DNA binding domains, *Science* 243, 1695, 1989.

48. Rauscher, F.J. III, Cohen, D.R., Curran, T., Bos, T.J., Vogt, P.K. et al., Fos-associated protein is the product of the jun proto-oncogene, *Science* 240, 1010, 1988a.

49. Rauscher, F.J. III, Sambucetti, L.C., Curran, T., Distel, R.J. and Spiegelman, B.M., A common DNA binding site for Fos protein complexes and transcription factor AP-1, *Cell* 52, 471, 1988b.

50. Rauscher, F.J. III, Voulalas, P.J., Franza, B.R. Jr. and Curran, T., Fos and Jun bind cooperatively to the AP-1 site: Reconstitution *in vitro*, *Gen. Dev.* 2, 1687, 1988c.

51. Sonnenberg, J.L., Rauscher, F.J. III, Morgan, J.I. and Curran, T., Regulation of proenkephalin by Fos and Jun, *Science* 246, 1622, 1989.

52. Carter, M.S. and Krause, J.E., Structure, expression, and some regulatory mechanisms of the rat preprotachykinin gene encoding substance P, Neurokinin A, Neuropeptide K, and Neuropeptide τ, *J. Neurosci.* 10, 2203, 1990.

53. Kislaukis, E. and Dobner, P.R., Mutually dependent response elements in the *cis*-regulatory region of the neurotensin/neuromedin N gene integrate enviromental stimuli in PC12 cells, *Neuron* 4, 783, 1990.

54. Naranjo, J.R., Mellström, B., Achaval, M. and Sassone-Corsi, P., Molecular pathways of pain: Fos/Jun-mediated activation of a noncanonical AP-1 site in the prodynorphin gene, *Neuron* 6, 607, 1991.

55. Monstein, H.J., Identification of an AP-1 transcription factor binding site within the human cholecystokinin (CCK) promoter, *NeuroReport* 4, 195, 1993.

56. Merchant, K.M., c-*fos* antisense oligonucleotide specifically attenuates haloperidol-induced increases in neurotensin/neuromedin N mRNA expression in rat dorsal striatum, *Mol. Cell Neurosci.* 5, 336. 1994.

57. Merchant, K.M., Dobie, D.J., Totzke, M., Aravagiri, M. and Dorsa, D.M., Effects of chronic haloperidol and clozapine treatment on neurotensin and c-*fos* mRNA in rat neostriatal subregions. *J. Pharmacol. Exp. Ther.* 271, 460, 1994.

58. Hope, B., Kosofsky, B., Hyman, S.E. and Nestler, E.J., Regulation of immediate-early gene expression and AP-1 binding in the rat nucleus accumbens by chronic cocaine, *Proc. Natl. Acad. Sci. U.S.A.* 89, 5764, 1992.

59. Iadarola, M.J., Chaung, E.J., Yeung, C.L., Hoo, Y., Silverthorn, M., Gu, J. and Draisci, G., Induction and suppression of protooncogenes in rat striatum after single or multiple treatments with cocaine or GBR-12909, *NINDA Res. Monogr.* 125, 181, 1993.

60. Rosen, J.B., Chaung, E. and Iadarola, M.J., Differential induction of Fos protein and a Fos-related antigen following acute and repeated cocaine administration, *Mol. Brain Res.* 25, 168, 1994.

61. Zhang, W.G., Pennypacker, K.R., Ye, H., Merchenthaler, I.J., Grimes, L., Iadarola, M.J. and Hong, J.S., A 35 kD fos-related antigen is co-localized with substance P and dynorphin in striatal neurons, *Brain Res* 577, 312, 1992.

62. Hope, B.T., Nye, H.E., Kelz, M.B., Self, D.W., Iadarola, M.J., Nakabeppu, Y., Duman, R.S. and Nestler, E.J., Induction of a long-lasting AP-1 complex composed of altered Fos-like proteins in brain by chronic cocaine and other chronic treatments, *Neuron* 13, 1235, 1994.

63. Pennypacker, K.R., Zhang, W.Q., Ye, H. and Hong, J.S., Apomorphine induction of AP-1 DNA binding in the rat striatum after dopamine depletion, *Mol. Brain Res.* 15, 151, 1992.

64. Nakabeppu, Y. and Nathans, D., A naturally occurring truncated form of FosB that inhibits Fos/Jun transcriptional activity, *Cell* 64, 751, 1991.

65. Mumberg, D., Lucibello, F.C., Schuermann, M. and Müller, R., Alternative splicing of *fos*B transcripts results in differentially expressed mRNAs encoding functionally antagonistic proteins, *Genes Dev.* 5, 1212, 1991.

66. Nakabeppu, Y., Oda, S. and Sekiguchi, M., Proliferative activation of quiescent Rat-1A cells by ΔFosB, *Mol. Cell. Biol.* 13, 4157, 1993.

67. Dobrzanski, P., Noguchi, T., Kovary, K., Rizzo, P., Lazo, P.S. and Bravo, R., Both products of the *fos*B gene, FosB and its short form, FosB/SF, are transcriptional activators in fibroblasts, *Mol. Cell. Biol.* 11, 5470, 1991.

68. Yen, J., Wisdom, R.M., Tratner, I. and Verma, I.M., An alternative spliced form of FosB is a negative regulator of transcriptional activation and transformation by Fos proteins, *Proc. Natl. Acad. Sci. U.S.A.* 88, 5077, 1991.

69. Doucet, J.P., Nakabeppu, Y., Bedard, P., Hope, B.J., Nestler, E.J, Jasmin, B.J., Chen, J.-S., Iadarola, M.J., St-Jean, M., Wigle, N., Blanchet, P., Grondin, R. and Robertson, G.S., Chronic alterations in dopaminergic neurotransmission produce a persistent elevation of ΔFosB-like protein(s) in the striatum, *Eur. J. Neurosci.*, 8, 365, 1996.

70. Burns, R.S., Chiueh, C.C., Markey, S.P., Ebert, M.H., Jacobowitz, D.M. and Kopin, I.J., A primate model of parkinsonism: selective destruction of dopaminergic neurons in the pars compacta of the substantia nigra by N-methyl-4-phenyl-1,2,3,6-tetrahydropyridine, *Proc. Natl. Acad. Sci. U.S.A.* 80, 4546, 1983.

71. Davis, G.C., Williams, A.C., Markey, S.P., Ebert, M.H., Caine, E.D., Reichert, C.M. and Kopin, I.J., Chronic parkinsonism secondary to intravenous injection of meperidine analogues, *Psychiatr. Res.* 1, 249, 1979.

72. Langston, J.W., Ballard, P., Tetrud, J.W. and Irwin, J., Chronic parkinsonism in humans due to a product of meperidine-analog synthesis, *Science* 219, 979, 1983.

73. Blanchet, P., Bedard, P.J., Britton, D.R. and Kebabian, J.W., Differential effect of selective D-1 and D-2 dopamine receptor agonists on levodopa-induced dyskinesia in 1-methyl-4-phenyl-1,2,3,6-tetrahydropyradine-exposed monkeys, *J. Pharmacol. Exp. Ther.* 267, 275, 1993.

74. Pan, H.S. and Walters, J.R., Unilateral lesion of the nigrostriatal pathway decreases the firing rate and alters the firing pattern of globus pallidus neurons in the rat, *Synapse* 2, 650, 1988.

75. Albin, R.L., Young, A.B. and Penney, J.B., The functional anatomy of basal ganglia disorders, *Trends Neurosci.* 12, 366, 1989.

76. Morelli, M. and Di Chiara, G., Agonist-induced homologous and heterologous sensitization to D-1 and D-2-dependent contraversive circling, *Eur. J. Pharmacol.* 141, 101, 1987.

77. Morelli, M., Fenu, S., Garau, L. and Di Chiara, G., Time and dose dependence of the "priming" of the expression of dopamine receptor supersensitivity, *Eur. J. Pharmacol.* 162, 329, 1989.

78. Robertson, G.S. and Morelli, M., ΔFosB participates in the mediation of priming, *Abstr. Soc. Neurosci.*, 21, 2078, 1995.

79. Bannon, M.J., Elliott, P.J. and Bunney, E.B., Striatal tachykinin biosynthesis: regulation of mRNA and peptide levels by dopamine agonists and antagonists, *Mol. Brain Res.* 3: 31, 1987.

80. Engber, T.M., Boldry, R.C., Kuo, S. and Chase, T.N., Dopaminergic modulation of striatal neuropeptides: differential effects of D_1 and D_2 receptor stimulation on somatostatin, neuropeptide Y, neurotensin, dynorphin and enkephalin, *Brain Res.* 581, 261.

81. Nemeroff, C.B., Neurotensin: perchance an endogenous neuroleptic? *Biol. Psychiatr.* 15, 283, 1980.
82. Helke, C.J., Krause, J.E., Mantyh, P.W., Couture, R. and Bannon, J., Diversity of mammalian tachykinin peptidergic neurons: multiple peptides, receptors, and regulatory peptides, *FASEB J.* 4, 1606, 1990.
83. Bronstein, D.M., Ye, H., Pennypacker, P.M., Hudson, P.M. and Hong, J.S., Role of a 35 kDa fos-related antigen (FRA) in the long-term induction of striatal dynorphin expression in the 6-hydroxydopamine lesioned rat, *Mol. Br. Res.*, 23, 191, 1994.
84. Chavkin, C., James, I.F. and Goldstein, A., Dynorphin is a specific endogenous ligand of the kappa opioid receptor, *Science* 215, 413, 1982.
85. Temple, A. and Zukin, R.S., Neuroanatomical patterns of the mu, delta and kappa opioid receptors of rat brain as determined by quantitative *in vitro* autoradiography, *Proc. Natl. Acad. Sci. U.S.A.* 84, 4308, 1987.
86. Mansour, A., Fox, C.A., Meng, H., Akil, H. and Watson, S.J., Kappa$_1$ receptor mRNA distribution in the rat CNS: comparison to kappa receptor binding and prodynorphin mRNA, *Mol. Cell. Neurosci.* 5, 124, 1994.
87. Steiner, H. and Gerfen, C.R., Cocaine-induced c-*fos* messenger RNA is inversely related to dynorphin expression in striatum, *J. Neurosci.* 13: 5066, 1993.
88. Steiner, H. and Gerfen, C.R., Dynorphin opioid inhibition of cocaine-induced, D1 dopamine receptor-mediated immediate-early gene expression in the striatum, *J. Comp. Neurol.* 353, 200, 1995.
89. Matsumoto, R.R., Brinsfield, K.H., Patrick, R.L. and Walker, J.M., Rotational behaviour mediated by dopaminergic and nondopaminergic mechanisms after intranigral microinjection of specific mu, delata and kappa opioid agonists, *J. Pharmacol. Exp. Ther.* 246, 196, 1988.
90. Thompson, L.A. and Walker, J.M., Inhibitory effects of the kappa opiate U50,488 in the substantia nigra pars reticulata, *Brain Res.* 517, 81, 1990.
91. Herrera-Marschitz, M., Christensson-Nylander, I., Sharp, T., Staines, W., Reid M., Hökfelt, T., Terenius, L. and Ungerstedt, U., Striato-nigral dynorphin and substance P pathways in the rat, *Exp. Brain Res.* 64, 193, 1986.
92. Matsuda, H., Hiyama, Y., Terasawa, K., Watanabe, H. and Matsumoto, K., Enhancement of rotational behaviour induced by repeated administration of SKF38393 in rats with unilateral nigrostriatal 6-OHDA-lesions, *Pharmacol. Biochem. Behav.* 42, 213, 1992.
93. Cole, R.L., Konradi, C., Douglass, J. and Hyman, S.E., Neuronal adaptation to amphetamine and dopamine: molecular mechanisms of prodynorphin gene regulation in rat striatum, *Neuron* 14, 813, 1995.

Chapter 7

PLASTICITY OF GAD GENE EXPRESSION IN ANIMAL MODELS OF MOVEMENT DISORDERS

Marie-Françoise Chesselet and Jill M. Delfs

CONTENTS

0-8493-8550-4/96/$0.00+$.50
© 1996 by CRC Press, Inc.

1. INTRODUCTION

1.1   GABA-ergic Neurons in the Basal Ganglia

The majority of neurons in the basal ganglia use the neurotransmitter gamma aminobutyric acid (GABA) and contain glutamic acid decarboxylase (GAD), the enzyme of GABA synthesis. GABA-ergic neurons include output neurons of the striatum, and of its main target areas, the pallidum (external segment, or globus pallidus, and internal segment or entopeduncular nucleus), and the substantia nigra pars reticulata (Figure 1).[1,2] In addition, the striatum contains a subpopulation of GABA-ergic interneurons.[3-5] It is therefore likely that changes in activity of GABA-ergic neurons would have profound effects on basal ganglia function. One focus of research in our laboratory is the analysis of changes in GABA-ergic transmission in animal models of neurodegenerative diseases and after pharmacological treatments that affect the basal ganglia.

1.2   Why Measure GAD mRNA?

An anatomical peculiarity of GABA distribution in the basal ganglia is that each region contains several distinct populations of GABA-ergic neurons. The striatum contains both efferent neurons and interneurons; the pallidum and substantia nigra contain the axon terminals of GABA-ergic input neurons and the cell bodies and dendrites of GABA-ergic output neurons.[1,2] Therefore, measurements of GABA levels and GAD activity in these regions cannot provide information on changes in GABA-ergic transmission in distinct neuronal populations. In contrast, the mRNA encoding GAD is present in cell bodies but not in axon terminals of GABA-ergic neurons.[2] Furthermore, single cell analysis of GAD mRNA expression with emulsion autoradiography allows for the quantification of relative mRNA levels in the cell bodies of distinct neuronal populations within one region.[6] Thus, analysis of GAD mRNA expression can provide unique information about identified GABA-ergic neurons in the basal ganglia.

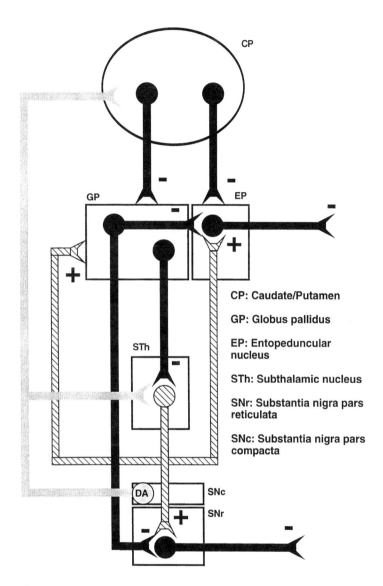

FIGURE 1.
Schematic diagram of basal ganglia circuitry. Black neurons are GABA-ergic; hatched neruons are glutamatergic and gray neurons are dopaminergic. Output from the striatum to the substantia nigra has been omitted for clarity.

Two forms of GAD exist in mammalian brain, GAD (M_r65,000: GAD65) and GAD (M_r67,000: GAD67).[7] These two GAD isoforms are encoded by distinct genes, and both mRNAs are co-expressed in all GABA-ergic neurons of the basal ganglia.[8] The level of mRNA encoding each isoform, however, varies among neuronal populations. The mRNA encoding GAD67 is much more abundant than GAD65 mRNA in striatal interneurons, whereas GAD65 is slightly more abundant than GAD67 mRNA in striatal efferent neurons.[8] The mRNAs encoding the two GAD isoforms are expressed at similar levels in neurons of the external pallidum and substantia nigra pars reticulata. In rats, GAD65 mRNA is much more abundant than GAD67 mRNA in the entopeduncular nucleus[8] but, curiously, this difference was not observed in primates, including humans (Chesselet et al., unpublished observation).

According to biochemical studies, GAD65 is not saturated with the cofactor pyridoxal phosphate, suggesting that this isoform may be predominantly regulated at the level of enzymatic activity.[7] In contrast, GAD67 is saturated with cofactor and is present at a higher level in cell bodies than in axon terminals of GABA-ergic neurons.[7,9] This led to the hypothesis that GAD67 may be predominantly regulated at the level of gene expression. In fact, changes in the level of GAD67 mRNA occur in parallel with changes in electrophysiological activity in GABA-ergic neurons in a number of experimental systems.[6,10-13] We have taken advantage of this property of GAD67 gene expression to obtain information about distinct populations of GABA-ergic neurons in the basal ganglia in animal models of movement disorders.

2. EFFECTS OF SHORT-TERM DOPAMINE DEPLETION AND DOPAMINE RECEPTOR BLOCKADE ON GAD67 mRNA IN THE STRIATUM

Unilateral injections of the neurotoxin 6-hydroxydopamine (6-OHDA) into the medial forebrain bundle just rostral to the substantia nigra after protection of noradrenergic neurons with the uptake blocker desipramine produce an extensive and selective destruction of nigrostriatal dopaminergic neurons.[14] This experimental model has been widely used as an animal model of Parkinson's disease, despite the obvious differences with the disease in which the loss of dopaminergic neurons is usually very progressive and bilateral.

2.1 Efferent Striatal Neurons

Previous studies from this laboratory show that unilateral dopamine depletion in adult rats induces a marked increase in GAD67 mRNA levels in the striatum ipsilateral to the lesion 2 and 3 weeks after surgery.[6] At these short post-lesion times, the effect is specific for GAD67 mRNA, without detectable changes in GAD65 mRNA, and is present in a large number of medium-sized striatal neurons, most likely efferent neurons. The increase in GAD67 mRNA is accompanied by an increase in immunoreactivity for the protein, GAD67.[6] These data confirmed the increase in striatal GAD67 mRNA observed by others using Northern blot analysis.[11,15] Previous studies show that dopaminergic lesions lead to an increase in GAD activity and in the spontaneous firing activity of striatal output neurons.[11,16] Taken together, these data suggest that the increase in GAD67 mRNA leads to an increase in functional GAD protein and parallels an increase in electrophysiological activity of the neurons. More recently, we have found that blockade of dopamine D2 receptors with haloperidol also increased GAD67 mRNA expression in the striatum. This effect was observed not only after short term (7 days) haloperidol treatment,[17] but also after administration of the drug for 8 months.[18]

2.2   Striatal GABA-ergic Interneurons

In contrast to observations in efferent striatal neurons, GAD67 mRNA levels were strikingly decreased in GABA-ergic interneurons after a unilateral 6-OHDA lesion in rats.[6] These GABA-ergic interneurons differ from striatal output neurons in that they express much higher levels of GAD67 mRNA.[3,6,8] In addition, these interneurons are unique among striatal neurons in that they express the calcium-binding protein parvalbumin and the Shaw-like potassium channel Kv3.1.[4,5,19] These molecular properties may contribute to the high firing rate of these neurons[20] and to their resistance to neurotoxic insults caused by injections of the NMDA agonist quinolinic acid or global ischemia.[21,22] Because of their high level of electrophysiological activity and their high content in GAD and GABA, it is likely that, despite their small number, these GABA-ergic interneurons contribute significantly to GABA release in the striatum.

The effects of decreased dopamine transmission on the firing activity of GABA-ergic interneurons is unknown. The selective decrease in GAD67 mRNA observed in these neurons, however, suggests that their activity may decrease after dopamine depletion. This could have important functional consequences because GABA-ergic interneurons of the striatum are thought to be responsible for IPSP following the initial EPSP evoked by cortical stimulation in striatal efferent neurons.[23] Thus a decreased activity in GABA-ergic interneurons after dopamine depletion could contribute to the increased activity observed in striatal output neurons.

2.3   Output Neurons of the Globus Pallidus

Striatal output neurons project to the two segments of the pallidum: external, or globus pallidus in rats, and internal, or entopeduncular nucleus in rats.[24] As previously mentioned, neuronal firing, GAD activity and GAD mRNA increase in striatum after dopamine depletion.[6,11,15,16] Furthermore, the mRNA encoding enkephalin, a neuropeptide co-localized with GABA in those striatal output neurons projecting to the globus pallidus, increase in striatal efferent neurons after dopaminergic lesions.[25,26] Finally, GABA release increases in the globus pallidus of these animals.[27] Taken together, these data suggest that after dopamine depletion, output neurons of the globus pallidus are inhibited by increased GABA-ergic input from the striatum.[24] In support of this hypothesis, the spontaneous firing of globus pallidus neurons decreases after 6-OHDA lesions in rats.[28] A similar effect was observed in the external pallidum of primates (the equivalent of the rat globus pallidus) after administration of 1-methyl-4-phenyl-1,2,3,6-tetrahydropyridine (MPTP), a toxin which induces the loss of nigrostriatal dopaminergic neurons in primates.[29] However, the levels of GAD mRNA were markedly increased in the globus pallidus of both rats and primates after short-term dopaminergic lesions.[26,30-33] Although it is possible that GAD mRNA levels may not always parallel the level of activity of GABA-ergic neurons, this result was surprising in view of the data indicating that, in many experimental conditions, GAD mRNA levels mirror changes in GABA-ergic transmission.[6,10-13]

One important observation that may explain the unexpected increase in GAD mRNA levels in the globus pallidus after dopaminergic lesions is that, in this condition, neurons of the globus pallidus show increased burst firing.[28,29] In other neurotransmitter systems, burst firing is associated with increased neurotransmitter release,[34] suggesting that GABA-ergic transmission may actually be increased rather than decreased in globus pallidus output neurons after dopamine depletion. Direct confirmation of this hypothesis is still lacking. However, haloperidol, a dopamine receptor antagonist which also increased GAD67 mRNA levels in the globus pallidus,[17] increases GABA release in the substantia nigra pars reticulata, a brain region which receives GABA-ergic inputs from the globus pallidus.[35]

The increase in GAD67 mRNA induced in the globus pallidus by haloperidol (1 mg/kg, s.c.) was observed after 3 days of haloperidol treatment, reached a maximum after 7 days, declined progressively at 14 and 21 days, and was no longer observed after 28 days of treatment.[17,36] To determine whether these changes were related to the behavioral effects of short-term haloperidol treatments, which include catalepsy in rats and akinesia in humans, rats were treated for 7 days with haloperidol alone or in combination with the muscarinic antagonist scopolamine (1 mg/kg, sc.). Co-administration of scopolamine with haloperidol blocked both haloperidol-induced catalepsy, and the increase in GAD67 mRNA in the globus pallidus (Tables 1 and 3).[17] In contrast, the haloperidol-induced increase in striatal enkephalin mRNA

TABLE 1.

GAD67 mRNA in Rat Pallidum after 7 Days of Haloperidol in the Absence or the Presence of Scopolamine

	dGP	vGP	EP	n
Controls	62.5 ± 2.3	60.7 ± 1.9	67.9 ± 4.7	5–7
Haloperidol	108.4 ± 7.3*	103.4 ± 6.8*	76.5 ± 2.2	6
Scopolamine	52.9 ± 1.4	53.7 ± 1.5	51.5 ± 2.9*	6–7
Haloperidol + Scopolamine	61.0 ± 2.1	61.4 ± 1.9	60.6 ± 2.5	7

Note: Rats were treated with haloperidol (1 mg/kg s.c.) alone or in the presence of scopolamine (1 mg/kg s.c.) for 7 days and euthanized 24 h after the last injection. *In situ* hybridization histochemistry for GAD67 mRNA was performed on sections of the globus pallidus (dorsal part: dGP and ventral part: vGP) and the entopeduncular nucleus. The autoradiographic signal was analyzed over single cells on emulsion-coated slides. Data are mean ±SEM of the average level of labeling in n rats. *p <0.05 with ANOVA followed by Dunnett's post-hoc test. (For details see Reference 17).

TABLE 2.

GAD67 mRNA in Rat Pallidum after Prolonged Haloperidol Treatment

	dGP	vGP	EP	n
28 Days				
Controls	87 ± 3	81.6 ± 2.9	82.0 ± 2.8	5–6
Haloperidol	89.9 ± 1.6	82.8 ± 1.4	107 ± 3.3*	5–6
8 Months				
Controls	77.6 ± 3.5	65.4 ± 3.1	65.9 ± 3	6
Haloperidol	59.7 ± 2.8*	54.6 ± 3.2*	83.4 ± 6.1*	6

Note: Rats were treated as described in details in Reference 18 and analysis was performed as for Table 1. *p <0.05 using ANOVA followed by Dunnett's post-hoc test.

TABLE 3.

Summary of Changes in GAD67 mRNA in the Globus Pallidus (GP) and Entopeduncular Nucleus (EP) after Haloperidol Treatment

Haloperidol	Behavior	GAD67 mRNA	
		GP	EP
7 Days	Catalepsy	↑	↔
7 Days+scopolamine	No catalepsy	↔	↔
28 Days	No catalepsy No oral dyskinesia	↔	↑
8 Months	Oral dyskinesia	↓	↑

levels was reduced but not abolished by scopolamine, suggesting that the antagonistic effect of scopolamine is not mediated solely by striatal projections to the globus pallidus.[17] A similar effect on striatal enkephalin mRNA was observed after administration of scopolamine to rats with 6-OHDA lesions of the substantia nigra.[37]

2.4 Output Neurons of the Entopeduncular Nucleus

Neurons of the entopeduncular nucleus constitute, along with those of the substantia nigra pars reticulata, the main output pathway of the basal ganglia to the thalamus and the reticular formation.[24,38] In addition to a direct input from the striatum, neurons of the entopeduncular nucleus are regulated by the striatum through connections with the globus pallidus and subthalamic nucleus (indirect pathway).[24,38] Entopeduncular neurons also receive a direct GABA-ergic projection from the globus pallidus.[39-41] A marked increase in the activity of neurons of the internal pallidum is observed after unilateral dopaminergic lesions in primates.[29,42] The resulting inhibition of thalamocortical neurons is believed to play a critical role in the pathophysiology of akinesia.[24,42-44]

In rats, the mRNAs encoding both isoforms of GAD were decreased in the contralateral entopeduncular nucleus after unilateral lesions of the nigrostriatal dopaminergic pathway.[30] In contrast, the levels of GAD67 or GAD65 mRNA were not affected in the entopeduncular nucleus ipsilateral to the lesion either 2 or 3 weeks post-surgery.[30] A clue to the lack of changes in GAD mRNA in the entopeduncular nucleus ipsilateral to the lesion may be provided by the time course of haloperidol effects in this region. After short term treatments with haloperidol (3 and 7 days, 1 mg/kg, s.c.), which induce an increase in GAD67 mRNA in the globus pallidus, no changes in GAD67 mRNA were detected in the entopeduncular nucleus.[17] However, GAD67 mRNA was increased in the entopeduncular nucleus in rats treated with the same dose of haloperidol for 14, 21 and 28 days.[17,18] A similar effect was found also in rats treated with haloperidol (1 mg/kg, i.p.) for 28 days[36] and in rats administered haloperidol for 8 months (Table 2).[18] Thus, the effect of haloperidol on GAD67 mRNA in the entopeduncular nucleus was only observed when the increase in GAD67 mRNA in the globus pallidus became smaller (14 and 21 days of treatment) or was no longer observed (28 days, 8 months of treatment).

The entopeduncular nucleus receives direct GABA-ergic inputs from the globus pallidus, as well as inputs from the striatum and the subthalamic nucleus.[38-41] Therefore, activation of GABA-ergic output neurons of the globus pallidus after short term haloperidol treatment, as suggested by the increase in GAD67 mRNA in this region, may oppose the combined effects of increased glutamatergic input from the subthalamic nucleus and decreased GABA-ergic input from the striatum on GAD mRNA expression in the entopeduncular nucleus. With longer haloperidol treatment, the effects on GABA-ergic output neurons in the globus pallidus decline, unmasking changes in GAD67 mRNA in the entopeduncular nucleus. It will be interesting to examine changes in GAD mRNA in the globus pallidus and entopeduncular nucleus at longer time points after nigrostriatal lesions to determine whether the same phenomenon occurs.

3. ROLE OF THE SUBTHALAMIC NUCLEUS IN THE REGULATION OF GAD mRNA IN THE BASAL GANGLIA

The subthalamic nucleus is a relatively small region, but it constitutes a main source of excitatory (glutamatergic) projections to the output regions of the basal ganglia, the pallidum and substantia nigra pars reticulata.[39,45] Lesions of the subthalamic nucleus have been shown to produce a "regularization" of the firing activity of neurons in the globus pallidus and entopeduncular nucleus in rats.[46] We have observed that subthalamic lesions produce a small decrease in the expression of GAD67 mRNA in the globus pallidus.[26] However, a similar effect was not seen in the entopeduncular nucleus at the time point (21 days post-lesion) examined. Interestingly, subthalamic lesions also decrease enzymatic markers of neuronal activity in the globus pallidus without affecting these parameters in the entopeduncular nucleus.[47] This further indicates that the globus pallidus and entopeduncular nucleus exhibit marked differences in their responses to perturbations in basal ganglia circuitry.

Although increased GABA-ergic control from the striatum is likely to contribute to the decrease in spontaneous firing activity observed in the globus pallidus after dopamine depletion,[28,29] it is unclear whether other inputs to the globus pallidus play a role in the increase in burst firing and GAD mRNA levels. The globus pallidus receives a significant glutamatergic projection from the subthalamic nucleus,[39] and the activity of subthalamic neurons is markedly increased in rats and primates with unilateral nigrostriatal lesions.[48,49] Therefore, we have tested the hypothesis that the subthalamic nucleus plays a role in the increase in GAD67 mRNA levels seen in the globus pallidus after 6-OHDA lesions.[30,31] Rats received a unilateral lesion of the subthalamic nucleus by local injection of a small volume (0.1 µl) of the neurotoxin, kainic acid 7 days before a unilateral 6-OHDA lesion of the dopaminergic nigrostriatal pathway on the same side. The increase in GAD mRNA levels seen in the globus pallidus ipsilateral to the dopaminergic lesion alone was completely abolished in rats with the combined lesions, suggesting that the subthalamic nucleus plays a critical role in this effect.[26] It remains unclear, however, whether this effect is mediated by the direct subthalamic-globus pallidus projection, or is indirectly mediated by polysynaptic pathways controlled by the subthalamic nucleus.[39] Indeed, changes in gene expression induced in the striatum by nigrostriatal lesions were also abolished in rats with the subthalamic lesions.[26] These data suggest that, in addition to the loss of dopaminergic projections to the striatum, other pathways affected by lesions of the nigrostriatal pathway play a critical role in the effects of dopamine depletion on striatal output neurons. This is in agreement with the observation that lesions of corticostriatal inputs also prevent the increase in enkephalin mRNA induced by 6-OHDA lesions.[50]

4. EFFECTS OF DOPAMINERGIC LESIONS ON GAD mRNA IN THE RETICULAR THALAMIC NUCLEUS

The reticular thalamic nucleus (RTN) is the only thalamic region that contains GABA-ergic neurons in rats.[2] GABA-ergic neurons of the RTN project to the dorsal thalamus and receive collaterals from thalamocortical and corticothalamic projections.[51] Because the basal ganglia project to the dorsal thalamus and affect activity in thalamocortical projections,[38] changes in the activity of basal ganglia neurons can alter activity in the RTN.[52] Furthermore, it was recently recognized that the RTN also receives direct GABA-ergic projections from the globus pallidus and substantia nigra pars reticulata.[53,54] Finally, the RTN receives inputs from the pedunculopontine tegmental nucleus, one of the main targets of the entopeduncular nucleus and substantia nigra pars reticulata.[55] Despite the existence of these multiple direct and indirect connections with the basal ganglia, little is known about the effects of perturbations in dopaminergic transmission on the activity of RTN neurons.

We have recently observed that unilateral lesions of the nigrostriatal dopaminergic pathway induce a bilateral increase in the expression of GAD67 mRNA in the ventral two thirds of the RTN.[56] No changes were observed in the dorsal part of the RTN, indicating a regional specificity for this effect (Figure 2). In contrast to dopaminergic lesions, blockade of D2-like dopaminergic receptors by either short or long term haloperidol treatment did not change GAD67 mRNA levels in the RTN. This suggests that the increase in GAD mRNA expression is not directly related to the changes in motor behavior induced by alteration in dopaminergic transmission. In contrast, this effect could play a role in

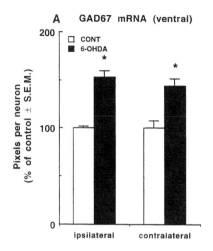
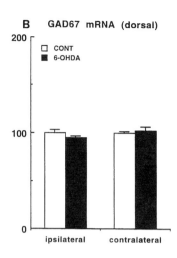

FIGURE 2.

Effects of unilateral 6-OHDA lesion of the nigrostriatal pathway on GAD67 mRNA in the ventral two thirds (ventral) and the dorsal one third (dorsal) reticular nucleus of the thalamus. Sections were processed for *in situ* hybridization histochemistry and the level of labeling over individual cells measured with an image analysis system. Data are the mean ± SEM of level of labeling (pixels occupied by siver grains) measured on the sides ipsilateral and contralateral to the lesion in control (cont) and lesioned (6-OHDA) animals, expressed as percent of the corresponding side of the controls. *p <0.05 when absolute values are compared to the corresponding side of controls with an unpaired two tailed Student t test. For details, see Reference 56.

other behavioral effects of dopamine lesions, such as deficits in cognition and attention, two behaviors which may be affected by the extensive connections of the RTN with the cerebral cortex,[57,58] and are altered after dopaminergic lesions [59,60]

5. EFFECTS OF LONG TERM TREATMENT WITH HALOPERIDOL ON GAD67 mRNA IN THE STRIATUM AND PALLIDUM

In patients with schizophrenia, long term administration of antipsychotic drugs can result in tardive dyskinesia, a severe side effect characterized by involuntary movements of the face, mouth and tongue.[61] The mechanisms underlying tardive dyskinesia are unknown and animal models of tardive dyskinesia remain controversial.[61,62] We have taken advantage of a drug regimen developed by Ellison and colleagues which produces oral dyskinesia in rats similar to tardive dyskinesia seen in humans after prolonged haloperidol treatment.[62] Rats were treated for 7 or 11 months with haloperidol decanoate, a slow release form of haloperidol, to produce a daily dose of 1 mg/kg, followed by administration of haloperidol in drinking water at the same dose for an additional 4 weeks to avoid confounding effects due to the slow clearance of haloperidol decanoate.[18,63] Rats were tested regularly for the presence of orofacial dyskinesia,[63] and only those animals displaying the abnormal movements were retained for the study.[18] Rats were sacrificed one week after cessation of the treatment.

In contrast to the increase in GAD67 mRNA levels seen in the globus pallidus of rats exhibiting catalepsy (short-term haloperidol), levels of GAD67 mRNA decreased in the pallidum of rats exhibiting oral dyskinesia (long-term haloperidol); (Table 2).[18] However, both short- and long-term haloperidol treatments increased enkephalin and GAD67 mRNA in the striatum, indicating that the effects in the striatum may not be related directly to the behavioral effects of the drug.[18] This further suggests that changes in GAD67 mRNA in the globus pallidus are related to changes in neuronal activity which play a critical role in the behavioral outcome of haloperidol treatment.

The mechanisms underlying the switch from increased to decreased expression of GAD67 mRNA in the globus pallidus after different duration of haloperidol treatment is unknown. Interestingly, however, the decrease in GAD67 mRNA observed in rats after long term haloperidol treatment is compatible with the observation that GAD activity decreases in the subthalamic nucleus, one of the main projection areas of the GABA-ergic neurons of the globus pallidus,[39] in humans with tardive dyskinesia.[64]

6. FUNCTIONAL IMPLICATIONS OF CHANGES IN GAD mRNA IN THE BASAL GANGLIA

Over the last few years, the results of numerous clinical, pathological and experimental studies have been integrated into a working model of basal ganglia function.[24] This heuristic model has led investigators to formulate hypotheses regarding changes in neuronal activity in relation to movement disorders.[24,42] A prominent feature of this model is the hypothesis that dopamine depletion in the basal ganglia leads to a decreased activity of globus pallidus output neurons projecting to the subthalamic nucleus. This decreased activity is believed to be the primary mechanism leading to the massive increase in activity of subthalamic neurons observed in rats and primates with lesions of the nigrostriatal pathway.[24,42] However, our knowledge of basal ganglia circuitry and electrophysiology has considerably evolved since the formulation of this commonly accepted model of basal ganglia circuitry. In parallel, clinical data have emerged showing that this model does not fully predict the spectacular outcome of surgical intervention in movement disorders of the basal ganglia.[44] Together with information from these studies, analysis of GAD gene expression in the basal ganglia after alteration of dopaminergic neurotransmission leads to some adjustments in the currently popular model of basal ganglia circuitry.

The increased level of GAD67 mRNA in the globus pallidus after dopamine depletion suggests that the increased excitatory response and burst firing activity observed in pallidal neurons may lead to increased rather than decreased GABA-ergic transmission in pallidal output neurons. Accordingly, activation of subthalamic neurons may be triggered by other mechanisms than an increased activity in output neurons from the globus pallidus. A possibility is a direct loss of dopaminergic stimulation of dopamine receptors within the subthalamic nucleus. This region receives collaterals from the nigrostriatal dopaminergic pathway and contains dopaminergic receptors.[65] Furthermore, local application of dopamine or dopaminergic agonists within the subthalamic nucleus induces changes in neuronal activity and elicits abnormal motor behavior in rats.[65] Alternatively, cortical inputs which make direct synaptic contacts with subthalamic output neurons could play a role in subthalamic activation after dopamine depletion.[38]

Taken together, the data on GAD gene expression in the basal ganglia in animal models of movement disorders point to a more central role of the external pallidum (globus pallidus) in the functional circuitry of the basal ganglia than previously thought. Unexpectedly, changes in GAD mRNA in this region correlate closely with changes in motor behavior, although the globus pallidus is viewed traditionally as a relay in the indirect pathway controlling the outputs of the basal ganglia through its projections to the subthalamic nucleus. Recent anatomical data, however, have shown that in addition to this pathway the globus pallidus sends direct projections to the internal pallidum, the substantia nigra pars reticulata and the reticular nucleus of the thalamus, also supporting a central role for the globus pallidus in basal ganglia circuitry.[39] These data also suggest that activation of the subthalamic nucleus after dopaminergic lesions may be mediated by different mechanisms than postulated in the current model of basal ganglia circuitry. Identifying the mechanisms responsible for this activation is essential in view of the dramatic effects of subthalamic lesions or inactivation by high frequency stimulation on the symptoms of Parkinson's disease.[66-68] In particular, recent data indicating that the subthalamic nucleus could be a target for at least some of the effects of dopaminergic,[65] serotonergic[69] and cholinergic[17] drugs on motor behavior suggest that neurotransmitter receptors in this region may become promising targets for the pharmacological treatment of movement disorders.

ACKNOWLEDGMENTS

We thank Ms. V.M. Ciaramitaro and Drs. G. Ellison, T.J. Parry and J.J. Soghomonian who contributed to some of the studies described in this chapter. We are also grateful to Dr. Allan Tobin, UCLA, for providing cDNAs and antibodies for these studies. Supported by PHS grants MH-44894 and training grant MH-17168.

REFERENCES

1. Smith, A. D. and Bolam, J. P., The neural network of the basal ganglia as revealed by the study of synaptic connections of identified neurones, *Trends Neurosci.*, 13, 259, 1990.
2. Chesselet, M.-F., Weiss, L. T., Wuenschell, C., Tobin, A.J. and Affolter, H.-U., Comparative distribution of mRNAs for glutamic acid decarboxylase, tyrosine hydroxylase, and tachykinins in the basal ganglia: an *in situ* hybridization study in the rodent brain, *J. Comp. Neurol.*, 262, 125, 1987.
3. Chesselet, M.-F. and Robbins, E., Characterization of striatal neurons expressing high levels of glutamic acid decarboxylase messenger RNA, *Brain Res.*, 492, 237, 1989.
4. Cowan, R. L., Wilson, C. J., Emson, P. C. and Heizmann, C. W., Parvalbumin-containing GABA-ergic interneurons in the rat neostriatum., *J. Comp. Neurol.*, 302, 197, 1990.
5. Kita, H., Kosaka, T. and Heizmann, C. W., Parvalbumin-immunoreactive neurons in the rat striatum: a light and electron microscopic study, *Brain Res.*, 536, 1, 1990.
6. Soghomonian, J.-J., Gonzales, C. and Chesselet, M.-F., Messenger RNAs encoding glutamate-decarboxylases are differentially affected by nigrostriatal lesions in subpopulations of striatal neurons, *Brain Res.*, 576, 68, 1992.
7. Erlander, M. G. and Tobin, A. J., The structural and functional heterogeneity of glutamic acid decarboxylase: a review, *Neurochem. Res.*, 16, 215, 1991.
8. Mercugliano, M., Soghomonian, J.-J., Qin, Y., Nguyen, H. Q., Feldblum, S., Erlander, M. G., Tobin, A. J. and Chesselet, M.-F., Comparative distribution of messenger RNAs encoding glutamic acid decarboxylases (Mr 65,000 and Mr 67,000) in the basal ganglia of the rat, *J. Comp. Neurol.*, 318, 245, 1992.
9. Gonzales, C., Kaufman, D. L., Tobin, A. J. and Chesselet, M.-F., Distribution of glutamic acid decarboxylase (Mr 67,000) in the basal ganglia of the rat: an immunohistochemical study with a selective cDNA-generated polyclonal antibody, *J. Neurocytol.*, 20, 953, 1991.
10. Litwak, J., Mercugliano, M., Chesselet, M.-F. and Oltmans, G., Increased glutamic acid decarboxylase (GAD) mRNA and GAD activity in cerebellar Purkinje cells following lesion-induced increases in cell firing, *Neurosci. Lett.*, 116, 179, 1990.
11. Segovia, J., Tillakaratne, N. J. K., Whelan, K., Tobin, A. J. and Gale, K., Parallel increases in striatal glutamic acid decarboxylase activity and mRNA levels in rats with lesions of the nigrostriatal pathway, *Brain Res.*, 529, 345, 1990.
12. Benson, D.L., Huntsman, M. M. and Jones, E. G., Activity dependent changes in GAD and preprotachykinin mRNAs in visual cortex of adult monkeys, *Cerebral Cortex*, 4, 40, 1994.
13. Martin, D. L. and Rimvall, K., Regulation of gamma-aminobutyric acid synthesis in the brain, *J. Neurochem.*, 60, 395, 1993.

14. Wooten, G. F. and Collins, R. C., Metabolic effects of unilateral lesion of the substantia nigra, *J. Neurosci.*, 1, 285, 1991.
15. Vernier, P., Julien, J. F., Rataboul, P., Fourrier, O., Feuerstein, C. and Mallet, J., Similar time course changes in striatal levels of glutamic acid decarboxylase and proenkephalin mRNA following dopaminergic deafferentation in the rat, *J. Neurochem.*, 51, 1375, 1988.
16. Schultz, W. and Ungerstedt, U., Short-term increase and long-term reversion of striatal cell activity after degeneration of the nigrostriatal dopamine system, *Exp. Brain Res.*, 33, 159, 1978.
17. Delfs, J. M., Anegawa, N. J. and Chesselet, M.-F., Glutamate decarboxylase messenger RNA in rat pallidum: comparison of the effects of haloperidol, clozapine and combined haloperidol-scopolamine treatments, *Neuroscience*, 66, 67, 1995.
18. Delfs, J. M., Ellison, G. D., Mercugliano, M. and Chesselet, M.-F., Expression of glutamic acid decarboxylase mRNA in striatum and pallidum in an animal model of tardive dyskinesia, *Exp. Neurol.*, 133, 175, 1995.
19. Lenz, S., Perney, T. M., Qin, Y., Robbins, E. and Chesselet, M.-F., GABA-ergic interneurons of the striatum express the Shaw-like potassium channel Kv3.1, *Synapse*, 18, 55, 1994.
20. Kawaguchi, Y., Physiological, morphological and histological characterization of three classes of interneurons in rat neostriatum, *J. Neurosci.*, 13, 4908, 1993.
21. Qin, Y., Soghomonian, J.-J. and Chesselet, M.-F., Effects of quinolinic acid on messenger RNAs encoding somatostatin and glutamic acid decarboxylase in the striatum of adult rats, *Exp. Neurol.*, 115, 200, 1992.
22. Gonzales, C., Lin, R.-C.-S. and Chesselet, M.-F., Relative sparing of GABA-ergic interneurons in the striatum of gerbils with ischemia-induced lesions, *Neurosci. Lett.*, 135, 53, 1992.
23. Kawaguchi, Y., Wilson, C. J., Augood, S. J. and Emson, P. C., Striatal interneurones: chemical, physiological and morphological characterization, *Trends Neurosci.*, 18, 527, 1995.
24. Albin, R. L., Young, A. B. and Penney, J. B., The functional anatomy of basal ganglia disorders, *Trends Neurosci.*, 12, 366, 1989.
25. Young, W. S., Bonner, T. I. and Brann, M., Mesencephalic neurons regulate the expression of neuropeptide mRNAs in the rat forebrain, *Proc. Natl. Acad. Sci. U.S.A.*, 83, 9827, 1986.
26. Delfs, J. M., Ciaramitaro, V. C., Parry, T. J. and Chesselet, M.-F., Subthalamic nucleus lesions: widespread effects on changes in gene expression induced by nigrostriatal dopamine depletion in rats, *J. Neurosci.*, 15, 6562, 1995.
27. Tossman, U., Segovia, J. and Ungerstedt, U., Extracellular levels of amino acids in striatum and globus pallidus of 6-hydroxydopamine-lesioned rats measured with microdialysis, *Acta Physiol. Scand.*, 127, 547, 1986.
28. Pan, H. S. and Walters, J. R., Unilateral lesion of the nigrostriatal pathway decreases the firing rate and alters the firing pattern of globus pallidus neurons in the rat, *Synapse*, 2, 650, 1988.
29. Filion, M. and Tremblay, L., Abnormal spontaneous activity of globus pallidus neurons in monkeys with MPTP-induced parkinsonism, *Brain Res.*, 547, 142, 1991.
30. Soghomonian J.-J. and Chesselet, M.-F., Effects of nigrostriatal lesions on the levels of messenger RNAs encoding two isoforms of glutamate decarboxylase in the globus pallidus and entopeduncular nucleus of the rat, *Synapse*, 11, 124, 1992.
31. Kincaid, A. E., Albin, R. L., Newman, S. W., Penney, J. B. and Young, A. B., 6-hydroxydopamine lesions of the nigrostriatal pathway alter the expression of glutamate decarboxylase messenger RNA in rat globus pallidus projection neurons, *Neuroscience*, 51, 705, 1992.

32. Soghomonian, J.-J., Pedneault, S., Audet, G. and Parent, A., Increased glutamic acid decarboxylase mRNA levels in the striatum and pallidum of MPTP-treated primates, *J. Neurosci.*, 14, 6256, 1994.
33. Herrero, M.-T., Ruberg, M., Hirsch, E. C., Guridi, J., Luquin, M. R., Guillen, J., Javoy-Agid, F., Agid, Y. and Obeso, J. A., Changes in GAD mRNA expression in neurons of the internal pallidum in parkinsonian monkeys after L-DOPA therapy, *Abstr. Soc. Neurosci.*, 19, 132, 1993.
34. Gonon, F. G., Nonlinear relationship between impulse flow and dopamine release by rat midbrain neurons as studied by *in vitro* electrochemistry, *Neuroscience*, 24, 19, 1988.
35. Yamamoto, B. K., Pehek, E. A. and Meltzer, H. Y., Brain region effects of clozapine on amino acid and monoamine transmission, *J. Clin. Psychiatr.*, 55, 8, 1994.
36. Mercugliano, M., Saller, C. F., Salama, A. I., U'Pritchard, D. C. and Chesselet, M.-F., Clozapine and Haloperidol have different effects on glutamic acid decarboxylase mRNA in the pallidal nuclei of the rat, *Neuropsychopharmacology*, 6, 179, 1992.
37. Nisenbaum, L. K., Kitai, S. T. and Gerfen, C. R., Dopaminergic and muscarinic regulation of striatal enkephalin and substance P messenger RNA following striatal dopamine denervation: effects of systemic and central administration of quinpirole and scopolamine, *Neuroscience*, 63, 435, 1994.
38. Parent, A. and Hazrati, L.-N., Functional anatomy of the basal ganglia. I. The cortico-basal ganglia-thalamo-cortical loop, *Brain Res. Rev.*, 20, 91, 1995.
39. Parent, A. and Hazrati, L.-N., Functional anatomy of the basal ganglia. II. The place of the subthalamic nucleus and external pallidum in basal ganglia circuitry, *Brain Res. Rev.*, 20, 128, 1995.
40. Hazrati, L.-N., Parent, A., Mitchell, S. and Haber, S. N., Evidence for interconnections between the two segments of the globus pallidus in primates: a PHA-L anterograde tracing study, *Brain Res.*, 533, 171, 1990.
41. Bolam, J. P. and Smith, Y., The striatum and the globus pallidus send convergent synaptic inputs onto single cells in the entopeduncular nucleus of the rat: a double anterograde labelling study combined with postembedding immunocytochemistry for GABA, *J. Comp. Neurol.*, 312, 456, 1992.
42. DeLong, M. R., Primate models of movement disorders of basal ganglia origin, *Trends Neurosci.*, 13, 282, 1990.
43. Ceballos-Baumann, A. O., Obeso, J. A., Vitek, J. L., DeLong, M. R., Bakay, R., Linazasoro, G. and Brooks, D. J., Restoration of thalamocortical activity after posterventral pallidotomy in Parkinson's disease, *Lancet*, 344, 814, 1994.
44. Marsden, C. D. and Obeso, J. A., The functions of the basal ganglia and the paradox of stereotaxic surgery in Parkinson's disease, *Brain*, 117, 877, 1994.
45. Kitai, S. T. and Kita, H., Anatomy and physiology of the subthalamic nucleus: a driving force of the basal ganglia, *Adv. Behav. Biol.*, 32, 357, 1987.
46. Ryan L. J. and Sanders, D. J., Subthalamic nucleus lesion regularizes firing patterns in globus pallidus and substantia nigra pars reticulata neurons in rats, *Brain Res.*, 626, 327, 1993.
47. Blandini, F. and Greenamyre, J. T., Effect of subthalamic nucleus lesion on mitochondrial enzyme activity in rat basal ganglia, *Brain Res.*, 669, 59, 1995.
48. Bergman, H., Wichmann, T., Karmon, B. and DeLong, M. R., The primate subthalamic nucleus. II. Neuronal activity in the MPTP model of parkinsonism, *J. Neurophysiol.*, 72, 507, 1994.
49. Robledo, P. and Feger, J., Acute monoaminergic depletion in the rat potentiates the excitatory effect of the subthalamic nucleus in the substantia nigra pars reticulata but not in the pallidal complex, *J. Neural. Trans.*, 86, 115, 1991.
50. Campbell, K. and Bjorklund, A., Prefrontal corticostriatal afferents maintain increased enkephalin gene expression in the dopamine-denervated rat striatum, *Eur. J. Neurosci.*, 6, 1371, 1994.

51. Jones, E. G., Some aspects of the organization of the thalamic reticular complex, *J. Comp. Neurol.*, 162, 285, 1975.

52. Lavin, A. and Grace, A. A., Modulation of the dorsal thalamic cell activity by the ventral pallidum: its role in the regulation of thalamocortical activity by the basal ganglia, *Synapse*, 18, 104, 1994.

53. Gandia, J. A., De Las Heras, S., Garcia, M. and Gimenez-Amaya, J. M., Afferent projections to the reticular thalamic nucleus from the globus pallidus and substantia nigra in the rat, *Brain Res. Bull.*, 32, 351, 1993.

54. Hazrati L. N. and Parent, A., Projections from the external pallidum to the reticular thalamic nucleus in the squirrel monkey, *Brain Res.*, 550, 142, 1991.

55. Hallanger, A. E., Levey, A. I., Lee, H. J., Rye, D. B. and Weiner, B. H., The origins of cholinergic and other subcortical afferents to the thalamus in the rat, *J. Comp. Neurol.*, 262, 105, 1987.

56. Delfs, J. M., Ciaramitaro, V. M., Soghomonian, J.-J. and Chesselet, M.-F., Unilateral nigrostriatal lesions induce a bilateral increase in glutamic acid decarboxylase mRNA in the reticular thalamic nucleus, *Neuroscience*, 71, 383, 1995.

57. Crick, F., Function of the thalamic reticular complex: the searchlight hypothesis, *Proc. Natl. Acad. Sci. U.S.A.*, 81, 4586, 1984.

58. Ross, D. T., Graham, D. I. and Adams, J. H., Selective loss of neurons from the thalamic reticular nucleus following severe human head injury, *J. Neurotrauma*, 10, 151, 1993.

59. Brown, R. G. and Marsden, C. D., Internal vs. external cues and the control of attention in Parkinson's disease, *Brain*, 111, 323, 1988.

60. Schneider, J. S. and Kovelowski, C. J., Chronic exposure to low doses of MPTP. I. Cognitive deficits in motor asymptomatic monkeys, *Brain Res.*, 519, 122, 1990.

61. See, R. S. and Chapman, M. A., The consequences of long-term antipsychotic drug administration on basal ganglia neuronal function in laboratory animals, *Crit. Rev. Neurobiol.*, 8, 85, 1994.

62. Ellison, G. D., Spontaneous orofacial movements in rodents induced by long-term neuroleptic drug administration: a second opinion, *Psychopharmacology*, 104, 404, 1993.

63. See, R. E. and Ellison, G. D., Comparison of chronic administration of haloperidol and the atypical neuroleptics, clozapine and raclopride, in an animal model of tardive dyskinesia in rat, *Psychopharmacology.*, 100, 404, 1990.

64. Andersson, U., Haggstrom, J. E., Levin, E. D., Bondesson, U., Valverius, M. and Gunne, L. M., Reduced glutamate decarboxylase activity in the subthalamic nucleus in patients with tardive dyskinesia. *Movement Disorders*, 4, 37, 1989.

65. Parry, T. J., Eberle-Wang, K., Lucki, I. and Chesselet, M.-F., Dopaminergic stimulation of the subthalamic nucleus elicits oral dyskinesia in rats, *Exp. Neurol.*, 128, 181, 1994.

66. Bergman, H., Wichmann, T. and DeLong, M. R., Reversal of experimental parkinsonism by lesions of the subthalamic nucleus, *Science*, 249, 1436, 1990.

67. Aziz, T. Z., Peggs, D., Sambrook, M. A., and Crossman, A. R., Lesion of the subthalamic nucleus for the alleviation of 1-methyl-4-phenyl-1,2,3,6-tetrahydropyridine (MPTP)-induced parkinsonism in the primate, *Movement Disorders*, 6, 288, 1991.

68. Limousin, P., Pollak, P., Bennazouz, A., Hoffman, D., LeBas, J.-F., Broussolle, E., Perret, J. E. and Benabib, A.-L., Effects on parkinsonian signs and symptoms of bilateral subthalamic nucleus stimulation, *Lancet*, 345, 91, 1995.

69. Eberle-Wang, K., Lucki, I. and Chesselet, M.-F., A role for the subthalamic nucleus in 5-HT2c-induced oral dyskinesia, *Neuroscience*, 72, 117, 1996.

Chapter **8**

PHARMACOLOGICAL AND MOLECULAR REGULATION OF THE EXPRESSION OF DOPAMINE RECEPTORS

Benjamin Weiss, Long-Wu Zhou, and Genoveva Davidkova

CONTENTS

0-8493-8550-4/96/$0.00+$.50
© 1996 by CRC Press, Inc.

1. INTRODUCTION

Dopamine is one of the major neurotransmitters in the basal ganglia and mesolimbic areas of the brain, and disturbances in dopaminergic activity have been implicated in the pathogenesis of frequently encountered motor and behavioral disorders. In particular, alterations in the density or sensitivity of D_2 dopamine receptors have been observed in schizophrenia, tardive dyskinesia, and Parkinson's disease.[1-5] Therefore, a more complete knowledge of the molecular regulation of dopamine receptors in the central nervous system, and the possible interactions that may exist between the dopaminergic system and other neurotransmitter systems may aid our understanding of important neuropsychiatric disorders, and perhaps identify novel targets for the selective therapeutic intervention of altered dopaminergic responses.

The interaction of neurotransmitters with their target receptor proteins results in two types of responses in the postsynaptic neurons: (1) acute activation of the receptors causing immediate electrophysiological and biochemical responses and (2) long-term alteration in the levels of the receptors, perhaps by modulating the expression of the genes encoding the receptors. The long-term modulation of receptors that takes place following chronic alterations in receptor activity is an important homeostatic mechanism by which an organism can adapt to changes in its environment. A better understanding of the homeostatic mechanisms governing the expression and activity of dopamine receptors might explain the chronic action of drugs acting on the dopamine system and the etiology of certain neuropsychiatric disorders. This brief review will focus on issues related to the long-term modulation of dopaminergic receptors with traditional pharmacologic agents and will consider novel molecular sites for the selective pharmacological alteration of dopaminergic responses using antisense oligodeoxynucleotides to alter the expression of specific subtypes of dopamine receptors.

2. GENERAL PRINCIPLES THAT GOVERN HOW RECEPTORS ARE MODULATED

The principles by which neuronal input modulates receptor function were initially characterized in detail for the β-adrenergic receptors in studies of the β-adrenergic receptor-adenylate cyclase complex of rat pineal gland.[6-12] This series of experiments showed that chronically increasing the input to the β-adrenergic receptors causes a decrease in the density and activity of postjunction β-adrenergic receptors, whereas chronically decreasing this input results in an increase in the density and function of these receptors. These studies have led to the now generally accepted hypothesis that the degree to which a receptor system can respond to stimulation is inversely related to the extent to which the receptors had been previously and persistently stimulated.[13] More recent studies provide abundant evidence that the principles of receptor modulation that were established for the β-adrenergic receptors, whereby the levels of receptors are controlled by a negative feedback mechanism, are of general biological significance and apply to receptors for other neurotransmitters as well, including the dopamine receptors.

3. MOLECULAR NEUROBIOLOGY OF DOPAMINE RECEPTORS

The major advances in the pharmacological and molecular regulation of the expression of dopamine receptors were facilitated by the rapidly increasing knowledge of the structure, pharmacology, second messenger coupling and anatomical localization of dopamine receptors.[14-16] Initially identified on the basis of pharmacological and biochemical properties,[17-19] the D_1 and D_2 dopamine receptor subtypes have now been classified into two major subfamilies on the basis of molecular cloning studies[15,16] (see also Chapter 3). Currently the D_1/D_5 receptor subfamily has two members which have high sequence homology in the transmembrane domains. This receptor subfamily exhibits the ligand-binding characteristics typical for the D_1 receptor: high affinity for benzazepines (SCH 23390, SKF 38393) and low affinity for butyrophenones (spiperone, haloperidol) and substituted benzamides (sulpiride) and is positively coupled to adenylate cyclase.

The D_2 receptor family is more complex, but the individual members also share high homology in the transmembrane domains. The D_2-like receptors have greater differences than the D_1-like receptors in their ligand-binding and second messenger coupling, and have not yet been completely characterized.[15,16] Among the different subtypes of dopamine receptors, the D_2[20] and particularly the D_3 and D_4 subtypes,[21,22] are thought to be involved in dopamine-mediated motor and behavioral dysfunction. Importantly, some classic D_2 dopamine receptor agonists and many antipsychotic drugs that in the past were believed to act only on the D_2 dopamine receptors have now been shown to bind also to the D_3 and D_4 dopamine receptors.[21,22]

4. BEHAVIORAL, BIOCHEMICAL AND MOLECULAR CORRELATES OF LONG-TERM ALTERATIONS IN DOPAMINERGIC INPUT

As could be predicted from studies of the β-adrenergic receptors, dopamine receptors should also be regulated through a negative feedback system, whereby excessive activation of the receptors leads to their down-regulation, and reduced input causes their up-regulation. Studies on the modulation of dopamine receptor expression have been greatly facilitated by the development of models for altering dopaminergic input and by the existence of compounds that interact relatively selectively with one of the subtypes of dopamine receptors, thereby leading to characteristic behavioral, biochemical and molecular changes.

4.1. 6-Hydroxydopamine Model of Dopaminergic Supersensitivity

Tardive dyskinesia and the positive symptoms of schizophrenia appear to be the result of excessive dopaminergic neurotransmission.[1-3,5] Denervation of the corpus striatum with the neurotoxin 6-hydroxydopamine and the resulting up-regulation of postsynaptic dopamine receptors has proved to be a useful model for studying the type of dopamine supersensitivity that may be related to some of the symptoms associated with these clinical disorders. Injection of 6-hydroxydopamine unilaterally into the striatum of mice, by destroying the dopaminergic nerve terminals, creates a denervation supersensitivity with the experimental animals rotating contralaterally to the lesioned side after acute challenge injections of D_1- or D_2-like dopamine agonists.[23-25] Using the 6-hydroxydopamine model in mice, the following questions on the regulation of D_1- and D_2-like receptors were addressed:

1. Can the behavioral supersensitivity be reversed (desensitized) by persistently administering high doses of dopaminergic agonists?
2. Is the desensitization selective for each subtype of dopamine receptor?
3. Are there differences in the mechanisms by which the D_1 and D_2 receptor subfamilies are regulated?
4. What are the resultant changes at the biochemical and molecular levels, particularly in the density of dopamine receptors and in the quantity of dopamine receptor mRNA, that are associated with altered dopaminergic behaviors?

The studies briefly summarized in the following section show that continuous exposure of dopaminergically supersensitive animals to D_1- and D_2-like dopamine receptor agonists, besides producing the expected acute actions on rotational behavior, resulted in long-term and differential effects on the levels and function of the dopamine receptor subtypes, providing experimental evidence that dopaminergic supersensitivity could be selectively altered.[14,26,27]

4.2   Chronic Treatment with Dopamine Agonists

4.2.1.   Behavioral Effects

Continuous infusion of dopaminergic drugs was achieved with subcutaneously implanted Alzet osmotic minipumps. In initial studies, mice with unilateral 6-hydroxydopamine lesions were treated continuously with the nonselective dopamine agonist apomorphine.[28] Prior to and within one hour of implantation of apomorphine, marked rotational behavior was observed. However, with continuous infusion of the drug the rotational behavior ceased. The absence of rotational behavior was due to the development of desensitization to apomorphine since the injection of animals with a challenge dose of apomorphine failed to produce rotational behavior. This desensitization to apomorphine was reversible upon removal of the pumps containing apomorphine.

Other studies showed that the dopamine supersensitivity in 6-hydroxydopamine lesioned mice can be *selectively* down-regulated by the continuous administration of agents selective for D_1- and D_2-like receptors.[29] This indicated that the D_1 and D_2 receptor systems can be desensitized independently, and that their long-term regulation is achieved by different mechanisms. Continuous infusions of the D_1-like agonists, SKF 38393, SKF 75670 and Cy 208-243, initially produced rotational behavior, which disappeared completely within 3 to 7 days of the drug infusion. In contrast continuous infusion of the D_2-like agonists, quinpirole and N-0437, also produced rotational behavior,

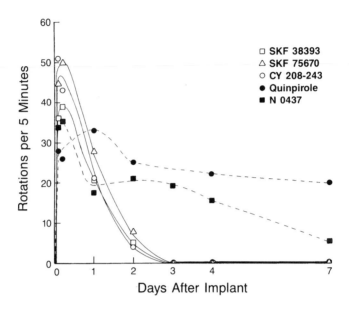

FIGURE 1.

Rotational behavior in supersensitive mice continuously exposed to D_1 and D_2 dopamine agonists. Mice with unilateral 6-hydroxydopamine lesions were implanted with Alzet mini pumps containing either SKF 38393 (16 μmol/kg/h), SKF 75670 (2 μmol/kg/h), CY 208-243 (5 μmol/kg/h), quinpirole (8 μmol/kg/h) or N-0437 (2.5 μmol/kg/h). Contralateral rotational behavior was determined during a 5-min period at the indicated times after implantation of each drug. Each point represents the means of five to seven animals. (From Weiss, B. et al. in *Neurochemical Phamacology — A Tribute to B. B. Brodie*, Costa, E., Ed., Raven Press, New York, 1989, 149. With permission.)

which although diminished, remained throughout the entire time of drug exposure (Figure 1). The selectivity of this response was demonstrated in lesioned mice continuously infused with CY 208-243. When rotations were no longer apparent, the mice failed to rotate to acute challenges with CY 208-243 or other structurally dissimilar D_1-like agonists; however, these mice still rotated to acute challenges with the D_2-like agonists, quinpirole and N-0437.

The long-term regulation of dopamine responses was characterized further by the continuous administration of D_1- and D_2-like agonists in normal mice. Many studies have previously shown that the acute administration of D_1 and D_2 agonists produce different behaviors: the D_1-like agonist, SKF 38393, in normal mice causes grooming, but no stereotypy, whereas the D_2-like agonist, quinpirole, causes a stereotypic response, but no grooming.[26,27] Continuous infusion of quinpirole with implanted Alzet minipumps initially produced stereotyped behavior, but this stereotypy was no longer apparent after several hours of infusion.[30] In an analogous experiment, continuously infusing SKF 38393

TABLE 1.

Effect of Long-Term Treatment of Mice with Selective D_1 and D_2 Agonists on D_1 and D_2 Receptors in Mouse Corpus Striatum

Treatment	6-OHDA-Lesioned Mice		Normal Mice	
	D_1	D_2	D_1	D_2
Continuous SKF 38393	88	92	n.d.	n.d.
Continuous quinpirole	99	64*	111	74*

Note: Mice with unilateral 6-hydroxydopamine-induced lesions of the corpus striatum were implanted with minipumps containing SKF 38393 (16 µmol/kg/h), quinpirole (8 µmol/kg/h) or vehicle. Normal mice were implanted with pumps containing quinpirole (2.5 µmol/kg/h) or vehicle. After 6 days the striata were removed and the binding of [³H] SCH 23390 (for D_1-like receptors) and [³H] spiroperidol for D_2-like receptors was determined. The results are expressed as a percentage of the binding in animals treated with vehicle. *p <0.01 when compared to vehicle-treated control; n.d. = not determined. Data taken from Weiss, B. et al.[14,27]

produced an initial grooming behavior which disappeared after several hours of infusion.[31]

These results showed that D_2 and D_1 family-mediated behaviors can be selectively down-regulated by the continuous infusion of D_2 and D_1 agonists in both normal and dopaminergically supersensitive mice. The down-regulation of supersensitive dopamine responses induced by continuous administration of D_2-like agonists might be of particular therapeutic significance since, by down-regulating D_2 dopamine responses, this treatment might prove useful for reversing the symptoms of tardive dyskinesia and the positive symptoms of schizophrenia.

4.2.2 Biochemical Effects

Studies of the biochemical correlates of altered dopaminergic behaviors showed that the decrease of D_2 subfamily-mediated behavioral responses after continuous administration of D_2-like agonists was associated with a reduction of D_2-like receptors measured by ³H-spiroperidol binding. These results were seen both in 6-hydroxydopamine-lesioned mice (Table 1) and in normal mice.[14,27] However, a discrepancy was still observed between the *extent* of reduction of receptors and behavior.[32] The possible mechanism by which behavior is altered to a greater extent than would be postulated from the slight changes in the levels of the receptors has been recently explained by the existence of a relatively small pool of rapidly turning over functional receptors.[33]

In contrast to the correlation that exists between D_2-like dopamine agonist-mediated behaviors and changes in the level of D_2-like dopamine receptors, there were no significant changes in the density of striatal D_1-like receptors after continuous administration of D_1-like agonists in

6-hydroxydopamine-lesioned mice[14] (Table 1). This apparent separation of alterations in behavior from alterations in the density of receptors has been reported also in studies in which other means were employed to change dopaminergic input, such as intermittently administering dopaminergic agonists to supersensitive animals[34,35] and lesioning with 6-hydroxydopamine.[36] It is possible that biochemical events beyond the receptor are responsible for the changes in behavior. Indeed, it has been demonstrated that modulation of dopamine receptor-mediated behaviors with agonists involves, besides alterations in the dopamine receptor levels themselves, a significant decrease in Ca^{2+}/calmodulin-dependent phosphorylation.[13] Another possible mechanism may involve a reduction in the number or sensitivity of D_1 receptors in sites outside of the striatum, such as in the substantia nigra, which contains D_1 receptors[37] and which may have a role in D_1-mediated rotational behavior (see also Chapter 4).

4.2.3 Molecular Effects

The distribution pattern of the D_2 receptor mRNA in mouse has been characterized previously in detail using *in situ* hybridization histochemistry.[27] The highest levels of D_2 receptor mRNA are found in the corpus striatum, olfactory tubercle and substantia nigra pars compacta; moderate levels are present in the nucleus accumbens and hypothalamus, and relatively low levels are seen in the cerebral cortex, septum, hippocampus, thalamus, cerebellum and pars reticulata. This distribution of D_2 mRNA correlates well with that of D_2-like receptors as determined from radioligand studies.

To uncover the molecular correlates of long-term D_1 and D_2 receptor regulation, the effects of chronic alteration of the input to dopamine receptors on the levels of the mRNAs encoding the dopamine receptors were studied using Northern blot and *in situ* hybridization histochemistry after continuous exposure of mice to D_1- or D_2-like agonists.[27,31] The results showed that down-regulation of D_2-mediated behavior and binding of D_2-like receptors after continuous administration of the D_2-like agonist, quinpirole, to mice is paralleled by a decrease in the levels of D_2 mRNA in the striatum (Figure 2). These data strongly suggest that there is a negative feedback regulation of D_2-mediated responses that involves gene expression. In contrast to the results with the D_2-like agonist, continuous treatment with the D_1-like dopamine receptor agonist, SKF 38393, failed to significantly change either the D_1 or D_2 receptor binding or the mRNAs for these receptors in the mouse striatum.[31] Our data are in agreement with the findings of others that chronic treatment with neuroleptics or dopaminergic denervation with 6-hydroxydopamine causes more consistent changes in D_2 receptors than in D_1 receptors.[38-42] Even though the biochemical and molecular

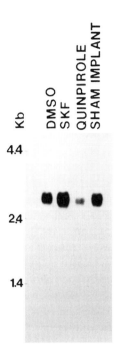

FIGURE 2.
Continuous treatment with dopamine agonists on D_2 dopamine receptor mRNA in the mouse striatum. Mice were implanted with pumps containing either the D_1 dopamine receptor agonist SKF 38393 (16 μmol/kg/h), the D_2 dopamine receptor agonist quinpirole (8 μmol/kg/h) or vehicle (12.5% ascorbic acid and 50% dimethylsulfoxide). After 7 days of treatment, animals were decapitated and the striata were removed. RNA was extracted on guanidine isothiocyanate-cesium chloride gradients and Northern blot analyses were performed. A single band of radioactivity is seen at 2.6 Kb which is identical to the molecular size of the D_2 dopamine receptor mRNA. The results show that continuous treatment with quinpirole, but not SKF 38393, decreases the level of the D_2 mRNA in the mouse striatum. (From Weiss, B. et al. *Adv. Biosci.*, 77, 9, 1990. With permission.)

basis for the modulation of D_1 receptors is still unclear, the existing data suggest that D_1 and D_2 receptors are regulated by different mechanisms and that D_2 receptors are more susceptible to pharmacological modulation than are D_1 dopamine receptors.

4.3 Persistent Blockade with Dopamine Receptor Antagonists

Another approach to study the long-term regulation of dopamine receptors is to decrease the dopaminergic input by the administration of dopamine antagonists. The studies based on this approach are of special importance for understanding the consequences of the therapy of diseases, such as schizophrenia, thought to be due to enhanced dopamine activity. Chronic blockade of D_2-like receptors with selective antagonists, as well as lesioning the nigrostriatal pathway with 6-hydroxydopamine, increased both the density of D_2-like receptors and D_2 receptor mRNA in striatum.[41-43] However, several other studies using chronic D_2-like antagonists failed to show significant changes in striatal D_2 receptor mRNA.[44,45] The observed discrepancies may be explained by differences in the magnitude and duration of action of the D_2 antagonists employed. For this reason the irreversibly acting D_2 antagonist fluphenazine-N-mustard (FNM) has proven particularly useful in evaluating the effects of continuously inhibiting D_2 receptors. In behavioral and biochemical studies from our laboratory, it was demonstrated that acute treatment with FNM is approximately 10 times more potent at inhibiting D_2- than D_1-subfamily mediated rotational behavior, and at inhibiting D_2- vs. D_1-like dopamine receptors in mice lesioned with 6-hydroxydopamine.[32,46] Similarly, repeated treatment of rats with FNM for 6 days produced almost complete inhibition of D_2-like receptors but only a relatively small inhibition of D_1-like receptors.[43] Repeated but not acute treatment with FNM resulted in statistically significant increases in D_2 receptor mRNA.[43] These results indicate that chronic blockade of D_2 dopaminergic input induces an up-regulation of D_2-like binding sites by increasing the expression of D_2 mRNA.

The changes that occur in the levels of D_1 receptors and their respective transcripts as a result of persistent blockade with dopamine antagonists are still unclear. In earlier studies, it was reported that chronic blockade of D_1-like receptors with antagonists or denervation with 6-hydroxydopamine increased D_1 receptor density.[38,47] However, other studies showed that lesioning the nigrostriatal pathway with 6-hydroxydopamine decreased the density of D_1 receptors as well as the levels of D_1 mRNA.[48-50] The changes in dopamine receptor density most likely result from alterations in the rate of receptor synthesis induced by changes in the levels of the respective receptor mRNAs. In support of this view are the data showing that repeated treatment with FNM or lesioning with 6-hydroxydopamine increases the rate of synthesis of D_2 receptors but decreases the rate of synthesis of D_1 receptors.[33,49,51]

5. INTERACTIONS BETWEEN VARIOUS NEUROTRANSMITTER SYSTEMS IN THE EXPRESSION OF DOPAMINE-MEDIATED BEHAVIORS

It is important to emphasize that psychomotor behaviors, which in the past have been thought to be regulated largely by dopamine neurotransmission, are now known to be the result of interactions of dopaminergic systems with several other neurotransmitter and neuromodulatory systems, including the glutamatergic, cholinergic and enkephalinergic systems.[52,53] Indeed, there is an increasing amount of evidence that the mRNAs for other neurotransmitter receptors change as a result of the continuous administration of pharmacologic agents acting on dopamine receptors. For example, lesions of the nigrostriatal dopaminergic pathway with 6-hydroxydopamine induced an increase in the level of glutamic acid decarboxylase (GAD) mRNA in the denervated striatum[54-57] and in the globus pallidus ipsilateral to the lesions[58] (see also Chapter 7). These changes in GAD mRNA could be reversed by chronically treating the animals with dopamine receptor agonists or by implanting dopaminergic grafts.[59,60] Recently it was shown that continuous administration of the irreversibly acting D_1- and D_2-like receptor antagonist, N-ethoxycarbonyl-2-ethoxy-1,2-dihydroquinoline (EEDQ), increases GAD mRNA in substantia nigra. Similarly, the D_2-like antagonist, FNM, increased GAD mRNA in the same brain region (Figure 3A).[61] These results suggest that changes of the levels of GAD mRNA induced by persistent blockade of dopamine receptors, particularly the D_2 subtype of dopamine receptors, may contribute to the motor dysfunctions induced by the long-term use of dopamine receptor antagonists.

Experimental evidence for molecular interactions between the dopaminergic and enkephalinergic systems was obtained after continuous administration of the D_2-like agonist, quinpirole, and the antagonist, FNM. Continuous treatment with quinpirole decreased the expression of proenkephalin mRNA and increased the levels of μ opioid receptors in the striatum.[31] By contrast, continuous treatment with FNM increased the expression of proenkephalin mRNA (Figure 3B) and reduced the density of μ and δ opioid receptors.[43] Finally, it was shown recently that continuous administration of D_4 dopamine receptor-preferring antagonist, clozapine, also increased the levels of proenkephalin mRNA.[62] These findings indicate that dopamine regulates the enkephalinergic system in striatum mainly via dopamine D_2-like receptors, a conclusion that is consistent with that suggested earlier by others.[63]

Interestingly, there is also evidence that *reciprocal* interactions exist between dopaminergic and other neurotransmitter systems. For example, blockade of NMDA receptors with dizocilpine (MK-801) alters certain dopaminergic behaviors.[64] Recent studies from our laboratory showed that long-term, but not acute, treatment with MK-801 caused a marked reduction in the levels of D_2 mRNA in mouse striatum (Figure 4), suggesting that glutamate plays an important regulatory role on the expression of D_2 dopamine receptors.[65]

Taken together, these results suggest that the molecular characterization of neurotransmitter systems that interact with the dopaminergic pathways in the CNS may aid our understanding of dopamine-mediated behaviors and offer new therapeutic approaches toward modifying abnormal dopaminergic neurotransmission.

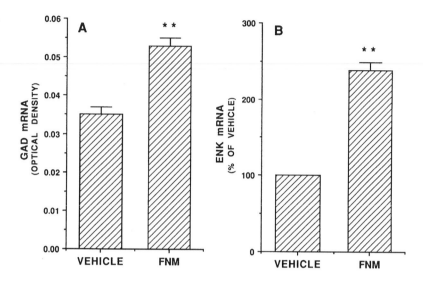

FIGURE 3.

Effects of FNM on glutamic acid decarboxylase mRNA in mouse substantia nigra and proenkephalin mRNA in rat striatum. (A) Mice were treated with FNM (20 μmol/kg, i.p.) or vehicle for 5 days and were killed 24 h after the last injection. Glutamic acid decarboxylase mRNA was determined by *in situ* hybridization histochemistry and quantified with the DUMAS image analyzer. Each bar represents the mean value from eight mice. (From Qin, Z.-H. and Weiss, B., *Eur. J. Pharmacol.*, 269, 25, 1994. With permission.) (B) Rats were treated with FNM (20 μmol/kg, i.p.) or vehicle for 6 days and were killed 20 h after the last injection. Proenkephalin mRNA was determined by *in situ* hybridization histochemistry and quantified with the DUMAS image analyzer. Each bar represents the mean value from 6 animals. Vertical brackets indicate the S.E. ** = $p < 0.01$ compared to the vehicle-treated group. (From Chen, J. F. et al., *Neurochem. Int.*, 25, 355, 1994. With permission.)

6. ANTISENSE OLIGODEOXYNUCLEOTIDES — SELECTIVE TOOLS FOR ALTERING THE EXPRESSION OF DOPAMINE RECEPTORS

A novel approach to pharmacologically alter the expression of dopamine receptors is the use of antisense oligodeoxynucleotides targeted to the mRNAs encoding the different receptor subtypes. Even though the classical dopamine agonists and antagonists have proven to be useful in activating or inhibiting the receptors, they possess several disadvantages. Many of the pharmacologic agents which were believed initially to be highly selective have been shown now to interact with more than one dopamine receptor subtype, thus leading to non-selective effects. These include the D_2 agonist, quinpirole and some of the most widely used antipsychotics.[21,22] Moreover, even selective

ANTERIOR STRIATUM   POSTERIOR STRIATUM

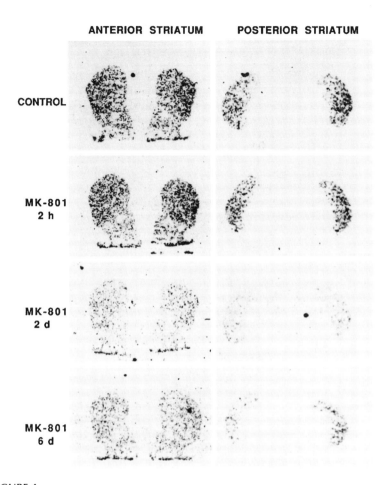

CONTROL

MK-801
2 h

MK-801
2 d

MK-801
6 d

FIGURE 4.
Effects of MK-801 on D_2 dopamine receptor mRNA in mouse brain. Mice were treated continuously with MK-801 (1.2 μmol/kg/h, s.c.) for 2 and 6 days using Alzet osmotic minipumps or were acutely injected with MK-801 (1.2 μmol/kg, i.p.) and were killed 2 h after the acute injection. D_2 dopamine receptor mRNA was determined by *in situ* hybridization histochemistry. (From Qin, Z.-H. et al., *Neuroscience*, 60, 97, 1994. With permission.)

actions on one subtype of receptor may result in effects at other subtypes, presumably because of an interaction between the dopamine receptor subtypes. For example, the D_1-like receptor antagonist, SCH 23390, not only inhibits grooming induced by the D_1-like agonist, SKF 38393, but also inhibits stereotypy and locomotor stimulation induced by D_2-like agonists.[36] Another disadvantage of the commonly used dopaminergic agents is that they induce long-term changes in the expression of dopamine receptors through the operation of negative feedback control mechanisms.[14] Furthermore, few selective ligands to dissect the biological functions of the newly discovered dopamine receptor subtypes (D_3, D_4, D_5) currently exist.[66-68]

During the past few years a major focus of our studies has been the design and *in vivo* application of antisense oligodeoxynucleotides targeted to the different subtypes of dopamine receptors, with the aims of understanding the regulation and function of the dopamine subtypes and of developing new, highly selective therapeutic agents.

A major difference between the traditional dopaminergic drugs and antisense oligodeoxynucleotides is that the latter are targeted to and interact with the nucleic acid coding for the receptor proteins. For the details on requirements and problems associated with the design of antisense oligodeoxynucleotides for *in vivo* application see Chapter 3 and other recent reviews on this subject.[69,70] Two very important issues, however, are the need for high specificity and stability of the oligodeoxynucleotides. The following paragraphs briefly summarize some of the studies on the application of antisense oligodeoxynucleotides, directed to the various dopamine receptor subtypes, as tools to modulate the expression and function of these subtypes of dopamine receptors.

6.1 Blockade of D_2 Dopamine Receptors with D_2 Antisense Oligodeoxynucleotides

In initial studies,[71] a 20-mer phosphorothioate oligodeoxynucleotide (D_2 antisense), targeted to a sequence of the D_2 dopamine receptor cDNA, bridging the initiation codon, was administered intracerebroventricularly (i.c.v.) into mice with unilateral 6-hydroxydopamine lesions of the corpus striatum. As a control, a 20-mer phosphorothioate oligodeoxynucleotide with the same proportion of bases as the D_2 antisense, but with a different sequence, was utilized. Administering the D_2 antisense caused an inhibition of quinpirole-induced rotational behavior, which was related to the dose and the number of injections of antisense oligomer. Similar i.c.v. injections of the random oligonucleotide or vehicle failed to inhibit quinpirole-induced rotational behavior. The effect of the D_2 antisense was reversible because 2 days after the cessation of antisense treatment, the quinpirole-induced rotational behavior reappeared.[19] Furthermore, the effects of the D_2 antisense were specific, because they blocked the actions of the D_2-like agonists, quinpirole and N-0437, but not those induced by the D_1 agonist, SKF 38393, or the muscarinic cholinergic agonist, oxotremorine (Figure 5). Moreover, the D_2 antisense inhibition of D_2-mediated rotational behavior could not be overcome by injecting higher doses of the D_2 agonist,

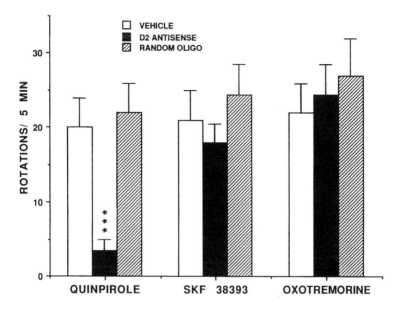

FIGURE 5.

D_2 antisense inhibits rotational behavior induced by quinpirole but not that induced by SKF 38393 or oxotremorine. Mice with unilateral intrastriatal lesions induced by 6-hydroxydopamine were administered intraventricular injections of vehicle (2 µl), D_2 antisense (2.5 nmol/2 µl) or random oligo (2.5 nmol/2 µl) three times at 12-h intervals. Mice were challenged with the D_2-like agonist, quinpirole (5 µmol/kg, s.c.), the D_1-like agonist, SKF 38393 (80 µmol/kg, s.c.) or the cholinergic muscarinic agonist, oxotremorine (5 µmol/kg, s.c.) 10 h after the last injection of vehicle or oligomer, and rotational behavior was assessed. Each point represents the mean value from five to seven mice. Vertical brackets indicate the S.E. *** = p <0.001 compared to vehicle- or random oligomer-treated mice. (From Weiss, B. et al., *Neuroscience*, 55, 607, 1993. With permission.)

quinpirole, suggesting that D_2 antisense treatment reduced the number of functional D_2 dopamine receptors. Other studies showing that D_2 antisense treatment of rats blocked the behavioral effects of the D_2 agonist, quinpirole, but not that produced by the D_1 agonist, SKF 38393, have supported our initial findings.[72]

At the biochemical and molecular levels, repeated i.c.v. treatment of mice with unilateral 6-hydroxydopamine lesions with the D_2 antisense significantly reduced the elevations of D_2-like binding sites and D_2 receptor mRNA in the dorsolateral striatum, but failed to significantly alter the D_1-like receptors and D_1 mRNA in the same brain area.[71] The relative selectivity by which the D_2 antisense appears to lower the *elevated* levels of D_2 mRNA to basal levels suggests the possibility that the induced mRNA may be more susceptible to inhibition by the antisense oligomer.

It should be emphasized that once again a discordance was observed between the relatively small reduction in D_2 receptors by the antisense oligomer and the profound changes in D_2-mediated behavior. Although the bases for this apparent discrepancy are still unclear, several possible explanations may be proposed. Perhaps it takes a long time for the total number of receptors to decrease, even if their synthesis is totally inhibited. This is supported by the data showing that the half life of the D_2 receptor is 3 to 4 days. Further, there may be different pools of D_2 receptors, not all of which are functional, and it may be necessary only to inhibit the synthesis of a relatively small pool of functional receptors in order to profoundly inhibit the dopamine-mediated effect. Finally, it is possible that the functional receptors responsible for the dopamine-mediated behavior turn over more rapidly than the nonfunctional receptors do. Evidence in support of these latter hypotheses was provided in experiments in which FNM was used to inhibit the total pool of D_2 receptors. Under these experimental conditions, treatment with D_2 antisense caused a marked reduction in the rate of recovery of D_2 receptors, which correlated well with the recovery of D_2 dopamine receptor-mediated function.[33]

6.2 Blockade of D_1 Dopamine Receptors with D_1 Antisense Oligodeoxynucleotides

The effect of i.c.v. administration of a 20-mer phosphorothioated oligodeoxynucleotide antisense to the D_1 dopamine receptor mRNA (D_1 antisense) on D_1-mediated behaviors was studied in normosensitive mice and in mice with dopamine supersensitivity induced by unilateral 6-hydroxydopamine lesioning of the corpus striatum. The D_1 antisense was designed to target a portion of the D_1 receptor cDNA bridging the initiation codon. A phosphorothioated random oligodeoxynucleotide was employed as a control.

Once again, the antisense treatment produced highly selective effects. The D_1 antisense inhibited grooming behavior induced by the D_1-like agonist SKF 38393 in normal mice, the reduction in grooming being related to the amount and length of time the D_1 antisense was given.[73] However, this treatment did not block the stereotypic effect induced by the D_2-like agonist quinpirole. Further, the D_1 antisense produced specific inhibitory effects on rotational behavior in dopaminergically supersensitive mice. As with the D_2 antisense treatment, the inhibitory effect of D1 antisense was specific in that it blocked the rotational effect of SKF 38393 but not that produced by quinpirole or oxotremorine (Figure 6). The D_1 antisense effect was also reversible, as there was a recovery from the inhibition of both grooming and rotational behaviors after cessation of D_1 antisense treatment.[73]

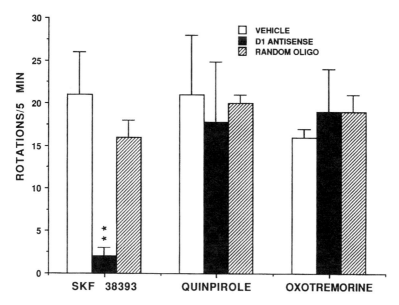

FIGURE 6.

D_1 antisense inhibits rotational behavior induced by SKF 38393 but not that induced by quinpirole or oxotremorine. Mice with unilateral intrastriatal lesions induced by 6-hydroxydopamine were administered intraventricular injections of vehicle (2 μl), D_1 antisense (2.5 nmol/2 μl) or random oligomer (2.5 nmol/2 μl) twice daily for two days. Mice were challenged with SKF 38393 (40 μmol/kg, s.c.), quinpirole (20 μmol/kg, s.c.) or oxotremorine (5 μmol/kg, s.c.) 10 h after the last injection of vehicle or oligomer, and rotational behavior was assessed. Each point represents the mean value from three to four mice. Vertical brackets indicate the S.E. ** = p <0.01 compared to vehicle- or random oligo-treated mice. (From Zhang, S.-P. et al., *J. Pharmacol. Exp. Ther.*, 271, 1462, 1994. With permission.)

7. CONCLUSIONS

1. Continuous blockade of dopaminergic input by selective antagonists or by lesioning the nigrostriatal pathway with 6-hydroxydopamine causes behavioral supersensitivity with concomitant increases in the density of D_2-like dopamine receptors and the levels of D_2 receptor mRNA. These data suggest that chronic reduction in dopaminergic input increases the rate of synthesis of D_2 dopamine receptors in the corpus striatum.

2. Continuous increases in D_2 dopaminergic input by administering selective D_2-like agonists is accompanied by a reduction in D_2-mediated behavior and a reduction in D_2-like binding sites as well as in D_2 mRNA. However, quantitatively the behavioral and molecular changes only partially correlate. Although continuous

treatment with D_1 agonists caused a reduction in D_1 dopamine receptor-mediated behavior, there was no corresponding reduction in the density of D_1 dopamine receptors.

3. Dopamine mediated behaviors and the expression of dopamine receptors in the CNS are the result of reciprocal interactions between several neurotransmitter systems, including the glutamatergic and enkephalinergic systems.

4. *In vivo* administration of oligodeoxynucleotides antisense to the D_2 or D_1 dopamine receptor mRNA causes selective and reversible blockade of D_2 and D_1 dopamine receptor-mediated behaviors, respectively. In contrast to that seen after continuous administration of traditional D_2-like antagonists, treatment with D_2 antisense produces a reduction in the levels of D_2 receptor density and D_2 mRNA, suggesting that D_2 antisense treatment may not lead to dopaminergic supersensitivity like that associated with chronically administered antipsychotic agents.

5. Dopamine receptor antisense oligodeoxynucleotides are novel pharmacological tools which should prove useful for probing the functions and properties of various subtypes of dopamine receptors. Studies using these agents already suggest that there is a small pool of rapidly turning over D_2 dopamine receptors that may constitute a functional pool of these receptors. These agents may provide also a novel and potentially more effective and selective means of treating disorders associated with dopaminergic supersensitivity.

REFERENCES

1. Seeman, P., Bzowei, N.H., Guan, H.-C., Bergeron, C., Reynolds, G.P., Bird, E.D., Riederer, P., Jellinger, K. and Tourtellotte, W.W., Human brain D1 and D2 dopamine receptors in schizophrenia, Alzheimers's, Parkinson's, and Huntington's Diseases, *Neuropsychopharmacology,* 1, 5, 1987.

2. Carlsson, A., The current status of the dopamine hypothesis of schizophrenia, *Neuropsychopharmacology,* 1, 179, 1988.

3. Ebadi, M. and Hama, Y., Dopamine, GABA, cholecystokinin and opioids in neuroleptic-induced tardive dyskinesia, *Neurosci. Biobehav. Rev.,* 12, 179, 1988.

4. Agid, Y., Cervera, P., Hirsch, E., Javoy-Agid, F., Lehericy, S., Raisman, R. and Ruberg, M., Biochemistry of Parkinson's disease 28 years later: a critical review, *Mov. Disord.,* 4, S126, 1989.

5. Carlsson, A. and Piercey, M., *Dopamine Receptor Subtypes in Neurological and Psychiatric Diseases, Clinical Neuropharmacology, Vol. 18, Suppl. 1,* Raven Press, New York, 1995, S215.

6. Weiss, B., Greenberg, L.H. and Blair Clark, M.B., Physiological and Pharmacological Modulation of the Beta-Adrenergic Receptor-Linked Adenylate Cyclase System: Supersensitivity and Subsensitivity, in *Dynamics of Neurotransmitter Function,* Israel Hanin, Ed., Raven Press, New York, 1984, 319.

7. Weiss, B. and Kidman, A., Neurobiological significance of cyclic 3′,5′-adenosine monophosphate, in *Advances in Biochemical Pharmacology,* Costa, E. and Greengard, P., Eds., Raven Press, New York, 1969, 131.

8. Weiss, B., Factors affecting adenyl cyclase activity and its sensitivity to biogenic amines, in *Biogenic Amines as Physiological Regulators,* Blum, J.J., Ed., Prentice-Hall, Inc., Englewood Cliffs, New Jersey, 1970, 35.

9. Weiss, B. and Crayton, J.W., Neural and hormonal regulation of pineal adenyl cyclase activity, in *Advances in Biochemical Psychopharmacology,* Greengard, P. and Costa, E., Eds., Raven Press, New York, 1970, 217.

10. Weiss, B., On the regulation of adenyl cyclase activity in the rat pineal gland, in *Cyclic AMP and Cell Function,* Robinson, G.A., Nahas, G.G. and Triner, L., Eds., *N.Y. Acad. Sci.,* New York, 1971, 507.

11. Weiss, B. and Strada, S., Neuroendocrine control of the cyclic AMP system of brain and pineal gland, in *Advances in Cyclic Nucleotide Research, Vol.1,* Greengard, P., Robinson, G.A. and Paoletti, R., Eds., Raven Press, New York, 1972, 357.

12. Weiss, B. and Strada, S., Adenosine 3′, 5′-monophosphate during fetal and postnatal development, in *Fetal Pharmacology,* Boreus, L., Ed., Raven Press, New York, 1973, 205.

13. Zhang, S.-P., Zhou, L.-W., Natsukari, N. and Weiss, B., Continuously infusing quinpirole decreases Ca^{2+}/calmodulin-dependent phosphorylation in mouse striatum, *Neurochem. Int.,* 23, 361, 1993.

14. Weiss, B., Cimino, M., Winkler, J.D. and Chen, J.F., Behavioral, biochemical, and molecular biological correlates of modulating dopaminergic responses, in *Neuro-Chemical Pharmacology — A Tribute to B. B. Brodie,* Costa, E., Ed., Raven Press, New York, 1989, 149.

15. Sibley, D., Monsma, F.J. and Shen, Y., Molecular neurobiology of D1 and D2 dopamine receptors, in *D1:D2 Dopamine Receptor Interactions,* Waddington, J.L., Ed., Academic Press, San Diego, 1993, 1.

16. Gingrich, J. and Caron, M.G., Recent advances in the molecular biology of dopamine receptors, *Annu. Rev. Neurosci.,* 16, 299, 1993.

17. Spano, P.F., Govoni, S. and Trabucci, M., Studies on the pharmacological properties of dopamine receptors in various areas of the central nervous system, *Adv. Biochem. Psychopharmacol.,* 19, 155, 1978.

18. Niznik, H.B., Dopamine receptors: molecular structure and function, *Mol. Cell. Endocrinol.,* 54, 1, 1987.

19. Kebabian, J.W. and Calne, D.B., Multiple receptors for dopamine, *Nature,* 277, 93, 1979.

20. Seeman, P., Brain dopamine receptors in schizophrenia and tardive dyskinesia, *Psychopharmacology,* Suppl. 2, 2, 1985.

21. Sokoloff, P., Giros, B., Martres, M.P., Bouthenet, M.L. and Schwartz, J.C., Molecular cloning and characterization of a novel dopamine receptor (D3) as a target for neuroleptics, *Nature,* 347, 146, 1990.

22. Van Tol, H.H., Bunzow, J.R., Guan, H.C., Sunahara, R.K., Seeman, P., Niznik, H.B. and Civelli, O., Cloning of the gene for a human dopamine D4 receptor with high affinity for the antipsychotic clozapine, *Nature,* 350, 610, 1991.

23. Ungerstedt, U. and Arbuthnott, G., Quantitative recording of rotational behavior in rats after 6-hydroxydopamine lesions of the nigrostriatal dopamine system, *Brain Res.,* 24, 485, 1970.

24. Ungerstedt, U., Postsynaptic supersensitivity after 6-hydroxydopamine induced degeneration of the nigrostriatal dopamine system, *Acta Psychiatr. Scand. Suppl.,* 367, 69, 1971.

25. Thornburg, J.E. and Moore, K.E., Supersensitivity to dopamine agonists following unilateral 6-hydroxydopamine-induced striatal lesions in mice, *J. Pharmacol. Exp. Ther.,* 192, 42, 1975.

26. Weiss, B., Goodale, D.B., Seyfried, D.M., Thermos, K. and Winkler, J., Modulation of dopaminergic behavioral responses, in *Modulation of Central and Peripheral Transmitter Function*, Biggio, G., Spano, P.F., Toffano, G.L. and Gessa, G.L., Eds., Liviana Press, Padova, 1986, 105.

27. Weiss, B., Zhou, L.-W., Chen, J.F., Szele, F. and Bai, G., Distribution and modulation of the D_2 dopamine receptor mRNA in mouse brain: Molecular and behavioral correlates, *Adv. Biosci.*, 77, 9, 1990.

28. Winkler, J.D. and Weiss, B., Reversal of supersensitive apomorphine-induced rotational behavior in mice by continuous exposure to apomorphine, *J. Pharmacol. Exp. Ther.*, 238, 242, 1986.

29. Winkler, J.D. and Weiss, B., Effect of continuous exposure to selective D1 and D2 dopaminergic agonists on rotational behavior in supersensitive mice, *J. Pharmacol. Exp. Ther.*, 249, 507, 1989.

30. Zhou, L.-W., Qin, Z.-H. and Weiss, B., Downregulation of stereotyped behavior and production of latent locomotor behaviors in mice treated continuously with quinpirole, *Neuropsychopharmacology*, 4, 47, 1991.

31. Chen, J.F., Aloyo, V.J. and Weiss, B., Continuous treatment with the D_2 dopamine receptor agonist quinpirole decreases D_2 dopamine receptors, D_2 dopamine receptor messenger RNA and proenkephalin messenger RNA, and increases mu opioid receptors in mouse striatum, *Neuroscience*, 54, 669, 1993.

32. Thermos, K., Winkler, J.D. and Weiss, B., Comparison of the effects of fluphenazine-N-mustard on dopamine binding sites and on behavior induced by apomorphine in supersensitive mice, *Neuropharmacology*, 26, 1473, 1987.

33. Qin, Z.-H., Zhou, L.-W., Zhang, S.-P., Wang, Y. and Weiss, B., D_2 dopamine receptor antisense oligodeoxynucleotide inhibits the synthesis of a functional pool of D_2 dopamine receptors, *Mol.Pharmacol.*, 48, 730, 1995.

34. Bevan, P., Repeated apomorphine treatment causes behavioral supersensitivity and dopamine D2 receptor hyposensitivity, *Neurosci. Lett.*, 35, 185, 1983.

35. Fayle, P., Jackson, D.M., Jenkins, O.F. and Lafferty, P.A., The effect of dopamine receptor agonist treatment on haloperidol-induced supersensitivity in mice, *Pharmacol. Biochem. Behav.*, 23, 715, 1985.

36. Breese, G.R. and Mueller, R.A., SCH-23390 antagonism of a D2-dopamine agonist depends upon catecholaminergic neurons, *Eur. J. Pharmacol.*, 113, 109, 1985.

37. Dubois, A., Savasta, M., Curet, O. and Scatton, B., Autoradiographic distribution of the D1 agonist [3H]SKF 38393 in the rat brain and spinal cord, *Neuroscience*, 19, 125, 1986.

38. Buonamici, M., Caccia, C., Carpentieri, M., Pegrassi, L., Rossi, A.C. and DiChiari, G., D2 receptor supersensitivity in rat striatum after unilateral 6-hydroxydopamine lesions, *Eur. J. Pharmacol.*, 26, 347, 1986.

39. Savasta, M., Dubois, A., Benavides, J. and Scatton, B., Different plasticity changes in rat striatal subregions following impairment of dopaminergic neurotransmission, *Neurosci. Lett.*, 85, 119, 1988.

40. Subramaniam, S., Lucki, I. and McGonigle, P., Effects of chronic treatment with selective agonists on the subtypes of dopamine receptors, *Brain Res.*, 571, 313, 1992.

41. Coirini, H., Schumacher, M., Angulo, J.A. and McEwen, B.S., Increase in striatal dopamine D2 receptor mRNA after lesions of haloperidol treatment, *Eur. J. Pharmacol.*, 186, 369, 1990.

42. De La Concha, A., McKie, J., Hodgkinson, S., Mankoo, B.S. and Gurling, M.H.D., Stereospecific effect of flupenthixol on neuroreceptor gene expression, *Mol. Brain Res.*, 10, 123, 1991.

43. Chen, J.F., Aloyo, V.J., Qin, Z.-H. and Weiss, B., Irreversible blockade of D_2 dopamine receptors by fluphenazine-*N*-mustard increases D_2 dopamine receptor mRNA and proenkephalin mRNA and decreases D_1 dopamine receptor mRNA and mu and delta opioid receptors in rat striatum, *Neurochem. Int.*, 25, 355, 1994.

44. Van Tol, H.H.M, Riva, M., Civelli, O. and Creese, I., Lack of effect of chronic dopamine receptor blockade on D2 dopamine receptor mRNA level, *Neurosci. Lett.*, 111, 303, 1990.

45. Srivastava, L.K., Morency, M.A., Bajwa, S.B. and Mishra, R.K., Effect of haloperidol on expression of dopamine D2 receptor mRNAs in rat brain, *J.Mol.Neurosci.*, 2, 155, 1990.

46. Winkler, J.D., Thermos, K. and Weiss, B., Differential effects of fluphenazine-N-mustard on calmodulin activity and on D_1 and D_2 dopaminergic responses, *Psychopharmacology*, 92, 285, 1987.

47. Creese, I. and Chen, A., Selective D1 dopamine receptor increase following chronic treatment with SCH 23390, *Eur.J.Pharmacol.*, 109, 127, 1985.

48. Joyce, J.N., Differential response of striatal dopamine and muscarinic cholinergic receptor subtypes to the loss of dopamine, *Exp.Neurol.*, 113, 261, 1991.

49. Qin, Z.-H., Chen, J.F. and Weiss, B., Lesions of mouse striatum induced by 6-hydroxydopamine differentially alter the density, rate of synthesis, and level of gene expression of D_1 and D_2 dopamine receptors, *J.Neurochem.*, 62, 411, 1994.

50. Marshall, J.F., Navarrete, R. and Joyce, J.N., Decreased striatal D1 binding density following mesotelencephalic 6-hydroxydopamine injections: an autoradiographic analysis, *Brain.Res.*, 493, 247, 1989.

51. Zhou, L.-W., Zhang, S.-P., Qin, Z.-H. and Weiss, B., *In vivo* administration of an oligodeoxyleotide antisense to the D_2 dopamine receptor mRNA inhibits D_2 dopamine receptor-mediated behavior and the expression of D_2 dopamine receptors in mouse striatum, *J.Pharmacol.Exp.Ther.*, 268, 1015, 1994.

52. Carlsson, M. and Carlsson, A., Interactions between glutamatergic and monoaminoergic systems within the basal ganglia-implications for schizophrenia and Parkinson's disease, *Trends Neurosci.*, 13, 272, 1990.

53. Graybiel, A.M., Neurotransmitters and neuromodulators in the basal ganglia, *Trends Neurosci.*, 13, 244, 1990.

54. Vernier, P., Julien, P., Rataboul, O., Fourrier, C., Feuerstein, C. and Mallet, J., Similar time-course changes in striatal levels of glutamic acid decarboxylase and proenkephalin mRNA following dopaminergic deafferentation in the rat, *J.Neurochem.*, 51, 1375, 1988.

55. Lindenfors, N.S., Brene, M., Herrera-Marshitz, M., Persson, H. and Sedvall, G., Region specific regulation of glutamic acid decarboxylase mRNA expression by dopamine neurons in rat brain, *Exp.Brain Res.*, 77, 611, 1989.

56. Segovia, J., Tillakaratne, N., Whelan, K., Tobin, A. and Gale, R., Parallel increases in striatal glutamic acid decarboxylase gene expression in host striatal neurons, *Brain Res.*, 529, 345, 1990.

57. Soghomonian, J.-J., Gonzalez, C. and Chesselet, M.-F., Messenger RNAs encoding glutamate decarboxylases are differentially affected by nigrostriatal lesions in subpopulations of striatal neurons, *Brain Res.*, 576, 68, 1992.

58. Kincaid, A.E., Albin, R.L., Newman, S.W., Penney, J.B. and Young, A.B., 6-Hydroxydopamine lesions of the nigrostriatal pathway alter the expression of glutamate decarboxylase messenger RNA in rat globus pallidus projection neurons, *Neuroscience*, 51, 705, 1992.

59. Segovia, J., Castro, R., Notario, V. and Gale, K., Transplants of fetal substantia nigra regulate glutamic acid decarboxylase gene expression in host striatal neurons, *Mol.Brain Res.*, 10, 359, 1991.

60. Savasta, M., Mennicken, F., Chritin, M., Abrous, D., Feuerstein, C., LeMoal, M. and Herman, J., Intrastriatal dopamine-rich implants reverse the changes in dopamine D2 receptor densities caused by 6-hydroxydopamine lesion of the nigrostriatal pathway, *Neuroscience*, 46, 729, 1992.

61. Qin, Z.-H. and Weiss, B., Dopamine receptor blockade increases dopamine D_2 receptor and glutamic acid decarboxylase mRNAs in mouse substantia nigra, *Eur.J.Pharmacol.Mol.Pharmacol.*, 269, 25, 1994.
62. Zhang, S.-P., Connell, T., Price, T., Simpson, G.M., Zhou, L.W. and Weiss, B., Continuous infusion of clozapine increases mu and delta opioid receptors and proenkephalin mRNA in mouse brain, *Biol.Psychiatr.*, 37, 496, 1995.
63. Tang, F., Costa, E. and Schwartz, J.P., Increase of proenkephalin mRNA and enkephalin content of rat striatum after daily injections of haloperidol for 2 to 3 weeks, *Proc.Natl.Acad.Sci.U.S.A.*, 80, 3841, 1983.
64. Liliequist, S., Ossowska, K., Grabowska-Anden, M. and Anden, N.-E., Effect of the NMDA receptor antagonist, MK-801, on locomotor activity and on the metabolism of dopamine in various brain areas of mice, *Eur.J.Pharmacol.*, 195, 55, 1991.
65. Qin, Z.-H., Zhou, L.-W. and Weiss, B., D2 dopamine receptor messenger RNA is altered to a greater extent by blockade of glutamate receptors than by blockade of dopamine receptors, *Neuroscience*, 60, 97, 1994.
66. Freedman, S., Patel, S., Marwood, R., Emms, R., Seabrook, G., Knowles, M.R. and McAllister, G., Expression and pharmacological characterization of the human D3 dopamine receptor, *J.Pharmacol.Exp.Ther.*, 268, 417, 1994.
67. Jarvie, K., Tiberi, M., Silvia, C., Gingrich, J.A. and Caron, M.G., Molecular cloning, stable expression and desensitization of the human dopamine D1B/D5 receptor, *J.Receptor Res.*, 14, 573, 1993.
68. Seeman, P., Guan, H.-C., Van Tol, H.M. and Niznik, H., Low density of dopamine D4 receptors in Parkinson's, schizophrenia and control brain striata, *Synapse*, 14, 247, 1993.
69. Weiss, B., Zhang, S.-P., and Zhov, L.-W., Antisense strategies in dopamine receptor pharmacology, *Life. Sci.*, in press.
70. Weiss, B., Davidkova, G., and Zhang, S.-P., Antisense strategies in neurobiology, *Neurochem. Int.*, in press.
71. Weiss, B., Zhou, L.-W., Zhang, S.-P. and Qin, Z.-H., Antisense oligodeoxynucleotide inhibits D_2 dopamine receptor-mediated behavior and D_2 messenger RNA, *Neuroscience*, 55, 607, 1993.
72. Zhang, M. and Creese, I., Antisense oligodeoxynucleotide reduces brain dopamine D_2 receptors: behavioral correlates, *Neurosci.Lett.*, 161, 223, 1993.
73. Zhang, S.-P., Zhou, L.-W. and Weiss, B., Oligodeoxynucleotide antisense to the D1 dopamine receptor mRNA inhibits D1 dopamine receptor-mediated behaviors in normal mice and in mice lesioned with 6-hydroxydopamine, *J.Pharmacol.Exp.Ther.*, 271, 1462, 1994.

INDEX

INDEX